PRACTICAL PHYSICS FOR ENGINEERS

AF631246

PRACTICAL PHYSICS FOR ENGINEERS

By

ASHOK KUMAR MISHRA

Lecturer Physics,
Amity University, Lucknow Campus,
Uttar Pradesh

UNIVERSITY SCIENCE PRESS
(An Imprint of Laxmi Publications Pvt. Ltd.)
An ISO 9001:2015 Company
BENGALURU • CHENNAI • GUWAHATI • HYDERABAD
KOCHI • KOLKATA • LUCKNOW • RANCHI
NEW DELHI

Practical Physics for Engineers

Copyright © by Laxmi Publications Pvt. Ltd.
All rights reserved including those of translation into other languages. In accordance with the Copyright (Amendment) Act, 2012, no part of this publication may be reproduced, stored in a retrieval system, translated into any other language or transmitted in any form or by any means—electronic, mechanical, photocopying, recording or otherwise. Any such act or scanning, uploading, and or electronic sharing of any part of this book without the permission of the publisher constitutes unlawful piracy and theft of the copyright holder's intellectual property. If you would like to use material from the book (other than for review purposes), prior written permission must be obtained from the publishers.

Printed and bound in India

First Edition : 2006, Edition : 2025

ISBN : 978-81-318-0747-7

Limits of Liability/Disclaimer of Warranty: The publisher and the author make no representation or warranties with respect to the accuracy or completeness of the contents of this work and specifically disclaim all warranties. The advice, strategies, and activities contained herein may not be suitable for every situation. In performing activities adult supervision must be sought. Likewise, common sense and care are essential to the conduct of any and all activities, whether described in this book or otherwise. Neither the publisher nor the author shall be liable or assumes any responsibility for any injuries or damages arising from here. The fact that an organization or Website if referred to in this work as a citation and/or a potential source of further information does not mean that the author or the publisher endorses the information the organization or Website may provide or recommendations it may make. Further, readers must be aware that the Internet Websites listed in this work may have changed or disappeared between when this work was written and when it is read.

All trademarks, logos or any other mark such as A-Cube, USP, Amanda, Golden Bells, Firewall Media, Triunity, Mathstick, Engpot, Laxmi appearing in this work are trademarks and intellectual property owned by or licensed to Laxmi Publications, its subsidiaries or affiliates. Notwithstanding this disclaimer, all other names and marks mentioned in this work are the trade names, trademarks or service marks of their respective owners.

Branches

Bengaluru	080-26 75 69 30
Chennai	044-24 34 47 26
Guwahati	0361-254
Hyderabad	040-27 55 53 83
Kochi	0484-405 13 03
Kolkata	033-40 04 77 79
Lucknow	0522-430
Ranchi	0651-224 24 64

PUBLISHED IN INDIA BY

UNIVERSITY SCIENCE PRESS
(An Imprint of Laxmi Publications Pvt. Ltd.)

An ISO 9001:2015 Company
113, GOLDEN HOUSE, GURUDWARA ROAD, DARYAGANJ,
NEW DELHI - 110002, INDIA
Telephone : 91-11-4353 2500, 4353 2501
www.laxmipublications.com
info@laxmipublications.com

PREFACE

There is a symbiotic relation between physics and engineering. Broadly speaking engineering is mainly Applied Physics. A proper study of Applied Physics is therefore indispensible for Engineering students.

This book is written with a view to provide a Guide to Engineering students of Amity University in Applied Physics Laboratory. All the experiments of I Semester and II Semester as per the syllabus are included. In addition, some more experiments have been included which are prescribed by UP Tech. and other Technical Institutions. I hope this book will be useful to first year students of Engineering Colleges.

I am thankful to Mr. Aseem Chauhan Member trustee (RBEF) for his constant encouragement and guidance.

I am also thankful to Maj. Gen. K.K. Ohri (Director, Amity Lucknow Campus) and Mr. Naresh Chandra (Dy. Director) for their invaluable supports in my academic development.

I express special thanks to Dr. (Prof.) H.M. Srivastava Head, Ph. D. Programes, (Amity Lucknow Campus) for his guidelines and significant suggestions.

I would like to express my gratitude to Prof. S.T.H. Abidi, former Prof. of Physics IIT, Khadagpur and Prof. Physics, Amity Lucknow Campus who inspired and encouraged me and also gave continuous support during the preparation of the manuscript.

Last but not least I am thankful to Laxmi Publications Pvt. Ltd. for the trouble they took in publishing this book in such a short time.

The author would be very much thankful to the readers for pointing out errors and omissions (if any) while reading the book so that it can be corrected in next edition.

Ashok Kumar Mishra
Lecturer Physics,
Amity University,
Lucknow Campus.
Tel. 2721 931, 2721 932 ext. 229
E-mail : akmishra_2k5@rediff mail.com.

Foreword

Prof. A.K. Mishra, Lecture Physics has written this practical book for Engineering students. Mr. Mishra has considered all possible points where students need help. His clear and simple presentation of experiments will help students to do the practicals independently. I am sure that this book will provide guide lines to teachers and students both.

Dr. H.M. Srivastava
B.Sc., M. Tech (IIT/B)
Ph. D. (IIT/B). FIE(I)
Ex. Principal KNIT, Sultanpur
Ex. Professor IIT, Bombay.
Head Ph. D. programmes of
Lucknow Campus, Amity University.

SYLLABUS

PHYSICS PRACTICALS

TAS-151/TAS-251

L T P
2 1 0

List of Experiments (Any Ten)

1. To determine the wavelength of monochromatic light by Newton's ring.
2. To determine the wavelength of monochromatic light with the help of Fresnel's biprism.
3. To determine the focal length of two lenses by nodal slide and locate the position of cardinal points.
4. To determine the specific rotation of cane sugar solution using half shade polarimeter.
5. To determine the wavelength of spectral lines using plane transmission grating.
6. To determine the specific resistance of the material of given wire using Carey Foster's bridge.
7. To determine the variation of magnetic field along the axis of a current carrying coil and then to estimate the radius of the coil.
8. To verify Stefan's Law by electrical method.
9. To calibrate the given ammeter and voltmeter.
10. To study the Hall effect and determine Hall coefficient, carrier density and mobility of a given semiconductor material using Hall-effect set up.
11. To determine energy bank gap of a given semiconductor material.
12. To determine E.C.E. of copper using Tangent or Halmholtz galvanometer.
13. To draw hysteresis curve of a given sample of ferromagnetic material and from this to determine magnetic susceptibility and permeability of the given specimen.
14. To determine the ballistic constant of a ballistic galvanometer.
15. To determine the viscosity of a liquid.

Note : Additional experiments may be added based on contents of syllabus.

CONTENTS

Dedicated

to

My Parents

1

Specific Rotation of Cane Sugar

EXPERIMENT

OBJECT

To determine the specific rotation of cane sugar solution.

APPARATUS REQUIRED

Polarimeter (with Laurent's half shade plate), sodium lamp, a glass tube of suitable length.

FORMULAE USED

The specific rotation α of the sugar solution is given by

$$\alpha = \frac{\theta v}{lx}$$

θ = rotation produced
l = length of the tube in decimeter
x = mass of the sugar dissolved
v = volume of the solution.

Basic principle involved. An unpolarized light after passing through a tourmaline crystal becomes polarised and vibrations are confined to only one plane which is perpendicular to the direction of propagation of light. The plane which contains vibrations is known as the plane of vibration. The plane $\perp$ to plane of vibration is called plane of polarisation. There are some substances through which when a plane polarized light is passed the plane of polarisation turns steadily about the direction of the beam. Such substances are known as optically active substances. The amount of rotation produced is directly proportional to the thickness of the optically active substance and inversely proportional to the square of the wavelength of the light. The former fact shows that the action occurs within the medium and not at the surface. When a plane polarized light passed through a medium containing several optically active substances, then the rotation produced is the algebraic sum of the individual rotation that would be produced by each of them. On looking towards the source of the light, if the optically active substance is found to rotate the plane of the polarisation to the right it is called *dextrorotatory*, if to the left, it is called *levorotatory*.

The specific rotation or rotating power of an optically active substance is defined as the angle of the rotation caused by :

1. 1 mm thick crystal if the substance is solid
2. 10 cms column
3. 10 cms column of solution containing 1 gm of the active substance for 1 cc of sol, in case the substance in a solution.

Specific rotation for an optically active solution. The amount of rotation θ of the plane of polarisation has been found proportional to (*i*) the length '*l*' of the column containing optically active substance, (*ii*) the concentration '*c*' of the solution.

Hence $\theta = \alpha\, l\, c = \alpha\, l\, x/v$

Where x is the mass of the substance dissolved in Vcc of a solvent, '*l*' is the length of the column in decimeter and 'α' is the constant of proportional known as the specific rotation of the solution. Thus

$$\alpha = \frac{\theta v}{l\, x}$$

by knowing the various quantities occurring in the above equation α, the specific rotation, can be evaluated. To determine the rotation of the plane of polarization, we employ polarimeter.

Laurents half shade polarimeter. The Laurent's half shade polarimeter consits of Nicole prisms N_1 and N_2. In between the polarizer n_1 and tube T. There is a half shade plate which is a circular plate made of glass, quartz combination. The other semicircular position ABC is a half wave plate of quartz and covers one half of the field of view. The other semicircular position ADC covering the other half of the field of view is made of glass and its thickness is adjusted in order to get same intensity of light transmitted by both positions.

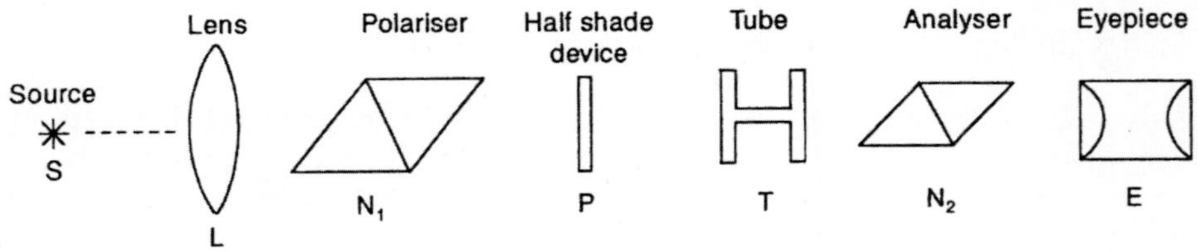

Fig. 1.1. Laurents half shade polarimeter.

Light beam from the monochromatic sources after passing through the lens 'L' and Nicole prism N_1 emerges as a plane polarized wave. This beam then passes through quartz and glass both. Let POP′ represent the plane of vibration of the plane polarized light emerging out from the polarizer and falling on the half shade. Let POP′ make an angle θ with optic axis YOY′. On resolving the incident vibration into two components OY (extraordinary) and OX (ordinary), the speed of ordinary component is greater than the extraordinary component within the quartz crystal and hence on emergence the ordinary component will gain a phase of π over the extraordinary component. Due to this change Π, the direction of the ordinary component is reversed and is represented by OX′. The resultant of OY and OX′ is in the direction QQ′. Therefore QQ′ is in the direction of vibration of the waves emerging out from glass. In view of this, the plane of polarization of light emerging out from glass at an angle 2θ. Thus the glass–quartz combination will provide two plane polarised beams having beams having vibrations parallel to OQ and OP respectively. If the principle section of the analog is along XOX′, the two halves on the field of view will have equal intensity since the position ON and ON′ as transmitted by the analyzer are equal. A slight rotation of the analyzer from this position in either of the two directions will cause one component greater than the other and due to this, the transmitted position ON and ON′ will not be equal and accordingly the intensity of one half would be different from the other half. Practically it has been found that in one position the quartz position appears brighter than the glass position whereas in some other position the reverse is seen. This variation in intensity has been found to be of great help for placing the analysis correctly for evaluating the amount of rotation.

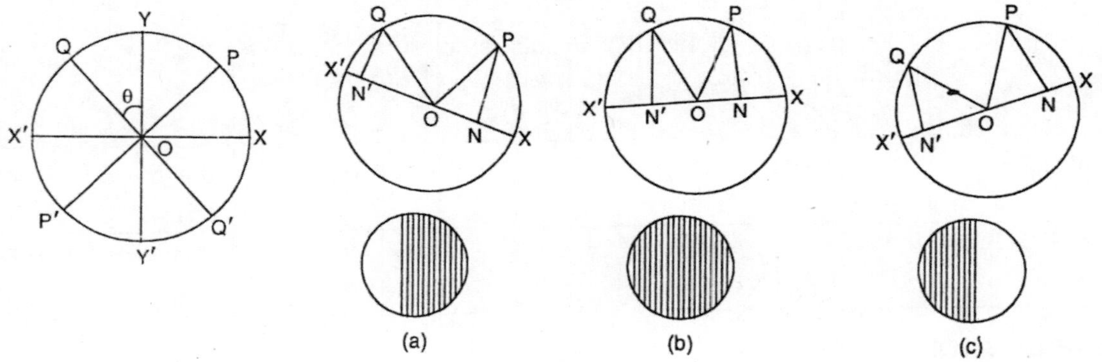

Fig. 1.2.

EXPERIMENTAL PROCEDURE

Weigh nearly 10 gms of sugar. Put it in 100 cc flask. Add some water and dissolve the sugar by shaking the flask gently. Fill the flask, unto mark scratched on it, by adding more water from outside. In this way the known mass of sugar is dissolved in 100 cc volume.

Take the polarimeter tube, clean it properly and fill it with water in such a way that no excess bubble remain inside it. Tight the tube with the end caps so that there should not be any leakage, put the tube in the place provided for it in the polarimeter. Switch on the source of the monochromatic light (sodium vapour lamp). On looking through the analyzer generally it will be observed that the two positions of the half shade will not have equal intensity, rotate the analyzer slowly and achieve that position of the analyzer at which the two halves of the field of view are equally intensifies. Note the reading of the scale attached with the polarimeter. Open one of its end caps and throw the water.

Rinse the tube with the help of prepared sugar solution and after that fill it carefully with the solution. Close this end with the help of the ends caps so that there should not be any leakage of water. Place the tube at its position provided and ensure that no air bubble remains inside.

Rotate the analyzer again till the two halves of the field of the view acquire equal intensities. Note this reading. This difference of the readings of the polarimeter with sugar solution and with water gives the angle of rotation θ.

OBSERVATIONS

Least count of analyser = ...

Mass of sugar dissolved x = ... gms

Volume of water used as solvent, v = ... cc

Length of polarimeter in decimeter, l = ... dcm

Temperature of the solution = ... °C

S.No.	*First position of analyzer*		*Position of analyser 180° apart from 1st position*		*Mean value of angle of rotation*
	Equal intensity of two position of the field of view obtained by :		*Equal intensity of two positions of the field of view obtained by :*		
	Clockwise rotation O′	*Anticlockwise rotation* O′	*Clockwise rotation* O′	*Anticlockwise rotation* O′	O′
WATER	a	b	C	d	$(a + b + c + d)/4 = \theta_1$
SOLUTION	a'	b'	C′	d'	$(a' + b' + c' + d)'/4 = \theta_2$

CALCULATION

$$\text{angle of rotation } \theta = \theta_1 \sim \theta_2$$

$$\alpha = \frac{\theta.v}{l.x} = \text{...... degree cc/gms/decimetres}$$

Probable errors. $\alpha = (\theta_1 - \theta_2)v/l.x$

Take the logarithm of both sides of those equation and differentiate it. One obtain for maximum probable error :

$$\frac{d\alpha}{\alpha} = \frac{d\theta_1 + d\theta_2}{\theta_1 - \theta_2} + \frac{dv}{v} + \frac{dl}{l} + \frac{dx}{x}$$

Nothing that $dv = 0$, $d\theta_1$ and $d\theta_2$ and least count of the polarimeter $dl = 0$, if length is given otherwise it represents the least count of the measurement and dx least count of the balance employment for weight calculated.

RESULT

The specific rotation of cane sugar at ....°C is α = ... degree cc/gm/dm

SOURCES OF ERROR AND PRECAUTIONS

1. The end cap should be air tight and there should be no air bubble inside the tube.
2. The polariser should not be disturbed throughout the course of the experiment.

3. It is important to take the observations when the position of the analyser is 180° apart from its initial position.

VIVA-VOCE

Q.1. What is specific rotation ?

Ans. $S_\lambda^t = \frac{\theta}{lc}$ where length l is in decimeter, c is conc. in gm/cc. and θ is rotation.

Thus, specific rotation of an optically active substance at a given temperature for a given wavelength is the rotation (in degrees) of plane of polarisation of incident polarised beam produced by one decimeter length of the substance of unit concentration.

Q.2. What are optically active substances ?

Ans. Optically active substance rotate the plane of polarisation or plane of vibration of a plane polarised light when it passes through it.

Q.3. What are the different optically active substances ?

Ans. (1) Right handed or *dextrorotatory* substances.

(2) Left handed or *Levorotatory* substances.

Q.4. What is meant be dextrorotatory and levorotatory substances ?

Ans. Dextrorotatory substances rotate the plane of polarisation of the incident polarised light in clockwise direction while levorotatory substances rotate in anti-clockwise direction as seen by the observer facing the emergent light.

Q.5. Give some examples of above two types of substances ?

Ans. Dextrorotatory–solution of camphor in alcohol, cane sugar, glucose etc. Levorotatory–solution fructose, nicotine, turpentine etc.

Q.6. What is the plane of vibration and plane of polarisation ?

Ans. *Plane of Vibration.* It is the plane containing the direction of propagation and the direction of vibration of light.

Plane of Polarisation. It is the plane passing through the direction of propagation and containing no vibration. This plane is always perpendicular to the plane of vibration.

Q.7. What is the effect of temperature of specific rotation ?

Ans. Specific rotation depends on temperature. In some substances it decreases with rise in temperature such as turpentine while in quartz, specific rotation increases.

Q.8. What is the unit of specific rotation ?

Ans. The unit is degree/decimeter/gm/cc.

Q.9. What is the practical application of the measurement of specific rotation ?

Ans. It is used for estimation of percentage of sugar in a given solution and also this method is used to determine the amount of sugar present in the urine of a diabetic patient.

Q.10. What is molecular rotation ?

Ans. Molecular rotation = specific rotation × molecular weight of the substance.

Q.11. What is a polarimeter ?

Ans. Polarimeter is used to measure the angle of rotation of plane of polarisation rotated by an optically active substance.

Q.12. What are the main parts of a polarimeter ?

Ans. Polariser, analyser and polarimeter tube.

Q.13. What is the function of polariser and analyser ?

Ans. Polariser changes an ordinary light into plane polarised light and analyser is used to analyse the light transmitted through the optically active substance and half shade on biquartz devices.

Q.14. What is a Nicol prism ?

Ans. Nicol prism in an optical device made from a calcite crystal for producing and analysing plane polarised light.

Q.15. What are ordinary and extraordinary light ?

Ans. Ordinary and extraordinary rays are plane polarised rays having vibration perpendicular to each other.

Ordinary rays obey the law of refraction but extraordinary rays do not follow the usual laws of refraction.

Q.16. What is the phenomena of double refraction ?

Ans. Double refraction is the splitting of an unpolarised beam when it passes through calcite or quartz crystal into ordinary and extraordinary rays.

Q.17. What is an optic axis ?

Ans. Optic axis is a direction in doubly refracting crystal along which all the plane waves move with the same velocity.

Q.18. What is the construction of half shade device ?

Ans. Half shade consists of two semicircular plates. Both the semicircular plates are cemented along the diameter. One semicircular plate is of glass and the other is of calcite crystal. The plate of calcite is half plate. Thickness of glass plate is adjusted such that it may absorb and reflect the same amount of light as the calcite plate.

Q.19. Why is the sodium light used in half shade polarisation ?

Ans. Half shade produces a path difference of $\frac{\lambda}{2}$ for a wavelength λ for which it is designed. Generally λ is matched with the sodium D line and hence we use sodium light.

Q.20. What is a bi-quartz device ?

Ans. It consists of two semicircular plates of left handed and right handed quartz which are cemented together to form a circular plate. Each of the quartz plate is cut perpendicular to the optic axis and their thickness is taken for which the angle of rotation for yellow colour is 90°.

Q.21. What are unpolarised, plain polarised and polarised light ?

Ans. The light having vibrations along all possible plane perpendicular to the direction of propagation is said to be unpolarised.

In case of polarised light, there is lack of symmetry about the direction of propagation.

The light is said to plain polarised if it is having vibrations only along a single direction, perpendicular to the direction of propagation.

2

Wavelength of Monochromatic Light Source by Newton's Ring

EXPERIMENT

OBJECT

To determine the wavelength (λ) of a monochromatic light source by Newton's ring experiment.

APPARATUS REQUIRED

Plano convex lens, glass-plate, travelling microscope, having main and vernier scale arrangement with stand. A glass plate with stand inclined at 45° to the horizontal, monochromatic light source (sodium lamp), speedometer.

FORMULAE USED

The wavelength (λ) of a monochromatic light source is given by :

$$\lambda = \frac{D_n{}^2 + p - D_n{}^2}{4pR}$$

Where

Dn = diameter of the nth dark ring

$Dn + p$ = diameter of $(n + p)$th dark ring

R = radius of curvature of Plano convex lens.

EXPERIMENTAL PROCEDURE

The ray diagram of experiment is shown in the Fig. 2.1 below :

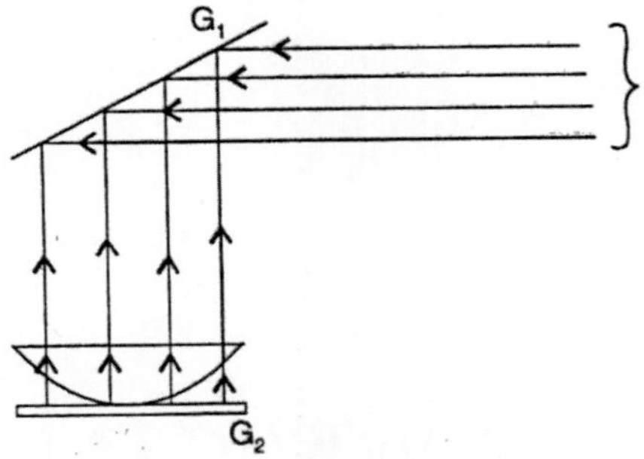

Fig. 2.1. Newton's ring experiment.

Follow the steps given below

1. Switch on the sodium lamp and wait until yellow light is emitted. This yellow light is allowed to fall on the glass plate G, which is inclined at 45° to the horizontal.
2. Place the convex lens over the glass plate g_2 so that an air film having variable thickness could be formed place this arrangement just below the glass plate g_1.
3. The light rays coming from source's gets reflected at plate g_1 and incident normally at the air film. Now focus the eyepiece of the travelling microscope.
4. Because of interference in thin film having variable thickness, circular rings are formed and can be seen by microscope by moving the eyepiece along the axis of main scale with the help of proper screw.
5 Adjust the centre of circular ring which is dark in reflected light and surrounded by alternate bright and dark rings in the field of view of eyepiece by moving the glass plate G_1 through stand. Make cross wire of eyepeices vertical and count the number of possible dark rings leaving central spot.
6. Fix the cross wire at the last ring (say 30th) to the left of the centre and note its position with the help of main vernier scale arrangement. Fix the cross wire at the next descending ring (29th ring) and note its position. Following the similar process for 28th, 27th, ... ring come upto 1st ring.
7. Again cross the central spot and fix the cross wire at 1st ring to the right of the centre and note its position with the help of main and vernier scale of microscope. Succeeding in similar way across 2nd, 3rd, ... rings, come to the 30th ring, right to the centre and note its position. Make the following observation table.

OBSERVATIONS

1. Determination of $(Dn + p)^2 - Dn^2$ and p.

   Least count of travelling microscope = ...

S. No.	Reading left to the centre (cms)	Reading right to the centre (cms)	Diameter	D_n^2	Let p = 15	Average of $D_n^2 + p - D_n^2$
	X_1	X_2	$D = X_1 - X_2$			

2. Determination of radius of curvature 'R' of Plano convex lenses.

$$R = \frac{l^2}{6h} + \frac{h}{2} \text{ in cm}$$

where l = distance between legs of sphenometer in cm.

h = spherometer reading.

VIVA-VOCE

Q.1. Why the light is red in the beginning when sodium lamp is switched on and then it becomes yellow ? Give the construction of sodium lamp.

Ans. Initially discharge pough neon gas and no discharge passes through sodium as its vapour pressure is low, then sodium atom is excited.

Sodium lamp consists of inverted U-shaped glass tube with 2 barium oxide coated tungsten electrodes. In U-shaped tube, sodium is deposited on the wall of the tube and neon gas is filled at a pressure of about 10 mm of mercury. The tube is enclosed in a vacuum jacket to minimize heat loss. Operating voltage of 440V is applied and discharge initially passes through neon gas and red colour is emitted. Now the voltage drops to 175V which is operating voltage for sodium. As soon as population of sodium increases, we get light of yellow colour. We get doublet of wavelength 5096 A, 5890A.

Q.2. What types of eye-piece are used in travelling microscope ?

Ans. Ramsden's eye-piece.

Q.3. What happens when white light is used instead of monochromatic light ?

Ans. Few coloured fringes will be observed near the centre.

Q.4. How can you get bright central spot in reflected light ?

Ans. By introducing sassafras oil between glass tube plate and Plano-convex lens because refractive index of material of glass plate and less than that of material of Plano-convex lens.

Q.5. What are Newton rings ?

Ans. When a Plano-convex lens is placed on a glass plate an air film of gradually increasing thickness is formed between glass plate and plano-convex lens. When monochromatic light is allowed to fall normally on air film and viewed in reflected light, alternate dark and bright rings formed are known as Newton's ring.

Q.6. How is Newton's ring formed ?

Ans. These rings are formed as a result of interference between the light waves reflected from the upper and lower surfaces of the air film developed between the convex surface of the Plano-convex lens and the plane glass plate.

Q.7. Why is the Newton's rings circular ?

Ans. They are circular because the air film formed is wedge shaped and loci of the points of equal thickness are circles concentric with the point of contact.

Q.8. Why the Plano-convex lens of large focal length is used ?

Ans. Because of large radius of curvature of Plano-convex lens, the angle of wedge is very small and rings observed have large diameter. So the error in measurements of diameter of the rings is minimized.

Q.9. What is the function of 45^0 inclined plane glass plate ?

Ans. The transparent glass plate inclined at 45^0 turns the light rays coming from an extended source to 90° and, so the rays falls normally on Plano-convex lens.

Q.10. Who do the rings get closer as the order of the rings increases ?

Ans. The direction of the dark rings are proportional to square root of the natural numbers ($dn\ \alpha'\ 2n - 1$) and the diameter of bright rings are proportional to the square root of odd natural nos. ($2\mu t(r + \theta) + \lambda/2$) thus the diameter of the rings does not increase in the same proportion as the order of the rings increases and as a result, the rings get closer and closer towards the end.

Q.11. Why is it necessary for the light to fall normally on the Plano convex lens ?

Ans. For normal incidence, the angle of incidence and reflection may be 0 so that the effective path difference between the interfering rays ($2\mu t(r + \theta) + \lambda/2$) in reflected system reduces to ($2\mu t + \lambda/2$) because cos $(r + \theta) = 1$.

Q.12. Why is the centre of the rings dark ?

Ans. At the point of contact of Plano convex lens and glass plate, the actual path difference between the two interfering rays is 0. But because one of the interfering rays is reflected from the glass plate which is in denser medium, the effective path difference is $\lambda/2$. Or the phase change of π. Thus at the centre, the condition of minimum intensity is satisfied and the central spot of the ring is dark.

Q.13. What will happen if the plane glass plate is replaced by a plane mirror ?

Ans. In this case, we will not get interference fringes between the intensity of light reflected from plane mirror will be so large that the interference fringes will not be visible and we will get uniform illumination.

Q.14. If sodium light is replaced by white light, which colour comes across first near the centre ?

Ans. We know that diameter of a ring depends on the wavelength of light used. So the rings of minimum wavelength will come first near the centre.

Q.15. Why do we use an extended source of light instead of a point source of a point source ?

Ans. Because of an extended source, we can view the whole air film.

Q.16. What will happen if a Plano convex lens of small radius of curvature is used ?

Ans. When the Plano convex lens of focal length is used the rings observed will be of small diameters and hence there will be error in measurments of their diameters.

Q.17. What will happen if Plano convex lens is replaced by a plane glass plate ?

Ans. In this case, interference takes place but the shape of the fringes will be irregular.

Q.18. What will happen if few drops of a transparent liquid are introduced between the Plano convex lens and the glass plate ?

Ans. The diameter of the rings decreases because diameter of ring is inversely proportional to $\sqrt{\mu}$ where μ is the refractive index of transparent liquid.

Q.19. What will happen if the Plano convex lens is replaced by a transparent glass plate inclined at a small angle with the lower glass plate ?

Ans. In this case bright and dark straight line fringes are observed in the field of view.

3

Moment of Inertia of Flywheel

EXPERIMENT

OBJECT

To determine moment of inertia of a flywheel about its own axis of rotation.

APPARATUS REQUIRED

Vernier Callipers, wheight box, thread, stopwatch and meter scale,

FORMULAE USED

The moment of inertia I of the flywheel J is given by

$$I = \frac{m\,gh\,t^2}{8\pi^2 n_2\,(n_1 + n_2)} - \frac{m\,r^2}{(1 + n_1/n_2)}$$

R = radius of the axle of the flywheel,

m = mass attached to the string,

l = length of the thread wrapped,

n_1 = number of turns the thread is wrapped,

n_2 = number of revolutions, fly-wheel makes before coming to rest.

Experimental set-up and basic principles involved. A flywheel is a heavy wheel having a long axle fixed on rigid frame with the help of two bearings. The axle contains either a hole or a peg in which one end of a thread may be fixed gently. A thread attached to a mass m at its one end is evently wrapped on the axle by its other end (see Fig. 3.1). Such an arrangement would not remain in static equilibrium because a downward force mg acts on the string. As the mass moves down, the thread unwind itself and looses its potential energy which is partly used in imparting the kinetic energy to the mass itself and partly in imparting the rotational kinetic energy to the flywheel. In addition to this, some energy would also be lost against the frictional forces. Ultimately, the mass and the thread detach from the flywheel. Let the linear velocity of the mass and the angular velocity of the flywheel at the instant of the detachment of the string be respectively v and ω. Then,

$$v = r\omega \qquad \ldots(1)$$

Where r is the radius of the axle of the flywheel. Let the mass fall through a distance has the string leaves the axle. The loss of potential energy of the mass is equal to mgh whereas the gain in its kinetic energy is equal to $\frac{1}{2}mv^2$ and the gain in rotational energy of the flywheel is equal to $\frac{1}{2}I\omega^2$, where I is the moment of inertia of the flywheel about its axle. Hence, applying the law of conversation of energy, we have

$$mgh = \frac{1}{2}mv^2 + \frac{1}{2}I\omega^2 + n_1F \qquad \text{...(2)}$$

where F represents the energy dissipated per unit revolution against the frictional forces and n_1 is the number of turns the thread is wrapped. When the thread is detached, the flywheel has its rotational energy $\frac{1}{2}I\omega^2$ and continues to rotate until it is dissipated against the frictional forces. If the flywheel makes n_2 revolution before it comes to rest, then,

$$\frac{1}{2}I\omega^2 = n_2F \qquad \text{...(3)}$$

Using this equation we may eliminate F from equation (2) and get

$$mgh = \frac{1}{2}mv^2 + \frac{1}{2}I\omega^2\left(I + \frac{n_1}{n_2}\right) \qquad \text{...(4)}$$

In view of equation (1), we write equation (4) as

$$1\left(1 + \frac{n_1}{n_2}\right) = 2mgh/\omega^2 - mr^2$$

The angular velocity ω may be determined by noting the time that flywheel takes to come to rest. Let it be t the flywheel in the beginning, when the thread leaves, rotates with angular velocity ω and after time t comes to rest. During this duration, the motion of the flywheel has been constantly and uniformly retarded. Hence we may consider the flywheel to the rotating for the time t with an average angular velocity $\frac{1}{2}(\omega + O) = \frac{1}{2}\omega$. Since the flywheel has made n_2 revolutions during this time, hence

$$\omega = 4\pi n_2/t$$

Substituting for ω from this equation in equation (1.4), one gets

$$I = \frac{mght_2}{8\pi^2 n_2(n_1 + n_2)} - \frac{mr^2}{1 + n_1/n_2}.$$

EXPERIMENTAL PROCEDURE

1. Take a thread of length smaller than the height of the axle of the flywheel above the floor of the laboratory. Insert one end of the thread which may be in the from of a pin, into the hole or the peg of the axle. Turn the flywheel slowly. So that the thread winds evenly on the axle without overlapping. Count the number of turns the thread has been wrapped. This is n.

2. Attach a known mass m at the other end of the thread and hold the flywheel with the support of hand. Let the flywheel, thread and the mass become stationary. Now leave the flywheel gently. The mass starts falling downward. In due course, the thread leaves the flywheel and the mass drops. At this instant, start the stopwatch and start counting the number of revolutions, wheel makes before coming to rest. The reading of the stopwatch gives t.
3. Measure the length of the thread wrapped on the axle. This may be verified by using the relation h-$2\pi r n_1$, where r is the radius of the axle. Measure the diameter of the axle with the help of Vernier Callipers in two perpendicular directions. Repeat the experiment with difference masses and take at least three observations for each mass. It would be convenient if n_1 is kept constant.

OBSERVATIONS

1. Measurement of h, n_1, n_2 and t

Least count of stop watch =

S. No.	M gm	n_1	n_2		t		h cm
			n_2	Mean n_2	Time t sec	Mean t sec	
1.							
2.							
3.							

2. Determination of the radius of the axle

Least count of the Vernier Callipers = ... cm

Zero error = + or – ...... cm.

S. No.	Reading along any diameter (a) cm.	Reading along perpendicular diameter (b) cm.	Mean diameter (a + b)/2 cm	Uncorrected mean diameter cm	Corrected mean diameter* cm	Mean radius cm
1.						
2.						
3.						
4.						
5.						

*Corrected mean diameter = Uncorrected mean diameter–Zero error

CALCULATION

$$I = \frac{mght^2}{8\pi^2 n_2(n_1 + n_2)} - \frac{mr^2}{(1 + n_1/n_2)}.$$

Calculate the first and second terms of the right hand side of this equation separately for any set of observations and show that the second term is negligibly small in comparison to the first. Hence neglect the second term for further calculation. Therefore, employ the formula

$$I = \frac{mght^2}{8\pi^2 n_2(n_1 + n_2)}$$

Calculate the moment of inertia for each set of observations and obtain the mean value of the moment of inertia.

Maximum Probable error

$$I = \frac{mght^2}{8\pi^2 n_2(n_1 + n_2)}$$

Take logarithm of both sides.

$\log I = \log m + \log g + \log h + 210\ gt - \log Tr - \log n_2 - \log (n_1 + n_2)$

Differentiate it to obtain

$$\frac{dI}{I} = \frac{dm}{m} + \frac{dg}{g} + \frac{dh}{h} + \frac{2dt}{t} - \frac{dn}{n_2} - \frac{d(n_1 + n_2)}{(n_1 + n_2)}$$

in view of our earlier discussion

$dm = dg = 0$; $dnj = 0$ and $dn_2 = 1$

dh = least count of metre scale

dt = least count of stop watch

as there will no be any error in counting n !, if there is any, it would be due to personal equation. Hence for maximum probable error.

$$\frac{dI}{I} = \frac{dh}{h} + \frac{2dt}{t} + \frac{1}{n_2} - \frac{1}{(n_1 + n_2)}$$

Calculate dI substituting the values of various quantities involved in this equation.

RESULT

The moment of inertia of the flywheel about its axis of rotation, I = $gm \times cm^2$.

Note. The significant figures of the result are decided in view of the probable error.

SOURCES OF ERROR AND PRECAUTIONS

1. The thread should be wrapped evenly over the axle of the flywheel and should not overlap.
2. The maximum error in the experiment occurs due to n_2 hence for the better result n_2 should be large. For this, m and h have to be kept large.
3. It should be checked that the stopwatch works well *i.e.*, it does not stop automatically after some time.

   It is convenient from the practical point of view to count $(n_1 + n_2)$ rather than n_2.

VIVA-VOCE

Q.1. Define Moment of inertia ?

Ans. (*a*) It is defined as numerically equal to the moment of couple or torque required to produce unit angular acceleration in the body ($\tau = id^2\theta/dt^2$)

(*b*) It is equal to the sum of the product of the mass to the square of the distance from the axis of rotation for every particle constituting the body *i.e.*, $1 = m_1r_1^2 + m_2r_2^2/ + ...$ where m_1, m_2 are the mass of the particles situated at respective distances r_1, r_2 ...... from the axis of rotation.

Q.2. What is its unit ?

Ans. In C.G.S. system : $gm \times cm^2$. In M.K.S. System : $kg \times m^2$.

Q.3. Is it constant for a body and if not, what are the factors on which it depends.

Ans. Moment of inertia of a body is not a constant. It depends on the axis of rotation. It depends on how the mass is distributed around axis of rotation.

Q.4. When is the moment of inertia of a body minimum and why ?

Ans. Moment of inertia of a body is minimum for the case when the axis of rotation passes through its centre of gravity because the moment of inertia, I of a body about any axis is equal to the sum of the moment of inertia I_g about a parallel axis passing through its centre of gravity and the product of its mass M to the square of the separation of the two axes, *i.e.*, $1 = I_g + ma^2$.

Q.5. What is radius of gyration ? Is it a constant ?

Ans. The radius of gyration, *k*, is the distance from the axis of rotation such that the mass of the body M when assumed to be situated at this distance will lead the moment of inertia, I, of the body *i.e.*, $I = Mk^2$.

Q.6. What is a flywheel and what are its uses ?

Ans. A flywheel is a heavy wheel having a long axle being fixed on rigid frame with the help of two bearings. It is used in locomotive engines, printing press and bicycle. The use of flywheel keeps this system in motion when its translatory motion ceases. Thus it conserves the motion.

Q.7. How would you increase the moment of inertia of a flywheel ?

Ans. We can make the moment of inertia large by arranging it more massive away from the axis of rotation. It may be achieved by use of spokes in the case of a flywheel.

Q.8. Why is a flywheel more concentrated at the rim ?

Ans. To have greater moment of inertia which causes to achieve greater rotational kinetic energy.

Q.9. What are the limitations and assumptions of the experiment ? Why is it necessary that thread should not overlap ?

Ans. It is assumed that the flywheel moves with the constant angular velocity before coming to rest. This is not rigorously correct. It is also assumed that the flywheel starts moving uniformly. In case thread overlaps this can not be achieved. In addition the thread should be wrapped evenly.

Q.10. What are the major sources of error ? How would you minimize them ?

Ans. It is evident from the expression for the probable error that the major error arises for the term $n_1 + n_2$ and n_2 in order to achieve more accuracy we must increase n_1 and n_2 to obtain this we may wrap longer thread and apply a heavier mass at the end of the thread.

4

Barpendulum

EXPERIMENT

OBJECT

To determine acceleration due to gravity (g) with the help of barpendulum and to study variation of (T) with distance (l).

APPARATUS REQUIRED

Bar pendulum and stopwatch.

FORMULA USED

Time period (T) of a compound pendulum is given by

$$T = 2\pi\sqrt{\frac{l + \frac{k^2}{l}}{g}}$$

$$g = \frac{4\pi^2}{T^2}\left(l + \frac{K^2}{l}\right)$$

Where l = distance of centre of suspension from the C.G. of the compound pendulum,
K = radius of gyration of the compound pendulum.

EXPERIMENTAL PROCEDURE

Follow the steps given below :

1. Suspended the bar pendulum at first hole which is 5 cms from the one end. Oscillate it in horizontal plane and measure the time of 20 oscillations with the help of a stopwatch. Note this is the first reading.
2. For subsequent reading suspend the pendulum at the 2nd, 3rd, 4th, ... 8th hole which are at 10 cm, 15 cm, 20 cm, ... 40 cms from one end respectively and measure the time of oscillations with the help of a stopwatch.

Q.2. Why bar pendulum becomes incapable to oscillate when it is suspended at 10th hole ?

Ans. Since $T = 2\pi$ if $l = 0$, then $T = \infty$ *i.e.*,

If compound pendulum is suspended at C.G., time period becomes infinite. Since barpendulum is a bar having length l m and contain 19 holes, each separated by 5 cm to next. At 10th hole, C.G. of bar lies and time period becomes infinite.

Q.3. How will you get radius of gyration of bar ?

Ans. Considering T v/s l graph and locate the point on L-axis. Corresponding to which it is minimum. Distance of this point from-axis from C.G. gives the value of K.

Q.4. Why time period at point S is name by same time period at point O and same for S′ and O′ ?

Ans. Because point of suspension and point of oscillation are interchangeable. S is point of suspension and O is point of oscillation and same for S′ as well as O′.

Q.5. In cricket game a batsman feels shock at his hand if ball could not strike at a particular point while batting ?

Ans. When we hold a cricket bat in our hands, the hands serve as an axis about which the bat can rotate. The centre of gravity is somewhere in the middle. Now taking the position of hands as the centre of suspension, there exists the corresponding centre of oscillation on the other side of the centre of gravity. If cricket ball strikes the bat at the centre of oscillation which will now be called centre of percussion, the bat simply rotates and there is no shock felt in the hands of the batsman.

5

Width of a Single Slit by Fraunhoffer Diffraction

EXPERIMENT

OBJECT

To determine the width of a single slit by producing diffraction pattern of fraunhoffer class with a monochromatic light using a spectrometer.

APPARATUS REQUIRED

Sodium vapour lamp, single slit affixed on a glass plate, spectrometer, prism, holder for mounting the slit, reading lens and reading lamp.

FORMULAE USED

The width 'b' of a single slit is given by

$$b = \frac{n\lambda}{\sin \theta_n}$$

Where

n = order of the minima and is equal to 1, 2, 3 etc. minima's

λ = wavelength of sodium light, and

θ_n = at which nth diffraction minima occurs.

Basic principle involved. Diffraction — it is a special class of interference in which the secondary wavelets originated from the different part of same wave front superimpose on each other and show a distribution of intensity which appears in form of bright and dark fringes in case of light.

EXPERIMENTAL PROCEDURE

1. Take a spectrometer and illuminate its slit with monochromatic source of light (sodium light). Reduce the width of the slit so that a fine pencil of ray may emerge out of collimator. Set the (*a*) collimator — to send parallel rays. (*b*) Telescope — to receive parallel rays and (*c*) plane of prism table horizontal.

2. Form a single slit by using two edges of shaving blade by way of finding it on one of the surfaces of a piece of transparent glass plate with the help of cellotape. Mount the glass plate in the grating holder and fix it on the prism table in such a way that the other surface of the glass plate faces the collimator and the plane of the slit be perpendicular to the rays incident. Increase the width of the slit so that the direct image of the slit of the spectrometer can be seen in the telescope. Coincide the direct image of the slit with the cross wire of the telescope and note the readings of two scales of spectrometer being placed at the angle of 180°, add 90° to this reading. Move the telescope and fix it at the total reading thus obtained. Rotate the prism slowly and look through the telescope. In this process, a reflected image of the slit with the cross-wire, the plate will make an angle of 45° with the incident beam. Now rotate the prism table through 45° in order to have a plane of the glass plate and hence the plane of the other slit, normal to the incident beams. In using, the relation $b \sin \theta_n = n\lambda$ for evaluating the width of the slit, this setting is essential. If one places the slit on the prism table in an arbitrary manner in that the angle of incidence assumes a finite value, about which he has no knowledge. The result evaluated in this manner with therefore is incorrect.
3. Put the telescope collinear with the collimator and reduce the width of the slit to such a small value at which the pattern seen in the telescope is identical to what has been displayed. Then the telescope on one side of the central maxima and fix the cross-wire on the fourth and fifth dark fringe which one is clearly visible and note the readings of the two verniers. Continue the process of rotating the telescope. In doing so you will come across the region which lies on the other side of the central maxima. Now fix the cross-wire on the 1st, 2nd, 3rd, minima respectively and note their positions with the help of two vernier scales. This process of taking the observations is to allow the telescope to move in one direction in order to avoid backlash error. The differene between the nth minima lying on the either side of the central maxima gives the values of $2\theta_n$. The half of this value is θ_n, the angle of diffraction nth minima. Once θ_n is known, 'b' can be calculated by using the relation $b \sin \theta_n = n\lambda$.

OBSERVATIONS

Least count of the spectrometer .....

Reading for direct image of the slit $\quad v_1 = \ldots\ldots \quad v_2 = \ldots\ldots$

Reading when telescope lie 90° ahead $\quad v_1' = \ldots\ldots \quad v_2' = \ldots\ldots$

Reading when telescope turned by back by 45° $\quad v_1'' = \ldots\ldots \quad v_2'' = \ldots\ldots$

(**Note** : $v_1' = v_1 + 45°$ and similar relations for the reading of the second vernier)

S. No.	Order of diffraction pattern	When diffraction pattern lies to				$2\theta_n$		Mean θ_n
		One side of principal maxima		One side of principal maxima		I Vernier	II Vernier	
		I Vernier X_1	II Vernier X_2	I Vernier X_1'	II Vernier X_2'	$X_1' - X_1$	$X_2' - X_2$	
1.	5							
2.	4							
3.	3							
4.	2							
5.	1							

CALCULATION

$$b = \frac{n\lambda}{\sin\theta_1} \quad \text{for first dark fringe} = l_1 \text{ cm}$$

$$b = \frac{n\lambda}{\sin\theta_2} \quad \text{for second dark fringe} = l_2 \text{ cm}$$

$$b = \frac{n\lambda}{\sin\theta_3} \quad \text{for third dark fringe} = l_3 \text{ cm}$$

Average value of b = 11 + 12 + 13 + ... + 15 cm = ... cm

Probable error. $b = \dfrac{n\lambda}{\sin\theta_n} \Rightarrow \boxed{\dfrac{db}{b} = \cot n\, d\theta_n}$ express $d\theta$ in radian.

Calculate db = ...

RESULT

The width of single slit obtained by diffraction pattern = ... cms.

SOURCES OF ERROR AND PRECAUTIONS

1. The mechanical and optical adjustment of the spectrometer should be made carefully.
2. While using the relation $b \sin\theta_n = n\lambda$ it is essential that the plane of the slit should be normal to incident beam.
3. Before taking any observation regarding the angle θ_n, it is important to ensure that the pattern is due to single slit and not due to its straight edges.
4. While taking observations regarding θ_n the telescope should be moved only in one direction in order to avoid backlash error.
5. While taking observations regarding θ_n one should fix the cross wire of the telescope on the minima not the maxima.

6

Dispersive Power of Prism

EXPERIMENT

OBJECT

To determine dispersive power (w) of material of the prism with the help of the prism and with the help of spectrometer.

APPARATUS REQUIERD

Spectrometer, prism, mercury lamps, magnifying lens, torch.

FORMULAE USED

Dispersive power (w) is given by

$$w = \frac{\mu_v - \mu_r}{\mu_y - 1}$$

where, μ_v = refractive index of violet colour.

μ_r = refractive index of red colour.

μ_y = refractive index of yellow colour.

$\approx (\mu_v + \mu_r)/2$

also $\mu = (\sin (A + \delta m)/2)/\sin A/2$

where A = prism angle

δm = minimum deviation angle.

Knowing the values of δm for violet, red, and yellow colour μv, μr and μy can be determined respectively.

EXPERIMENTAL PROCEDURE

(A) for the prism angle (A)

Follow the following steps given below :

1. Switch on mercury lamp. Focus the telescope of the spectrometer for distant object. Adjust the slit of the collimator to have a very small slit width and observe its sharp image with help of telescope, by moving collimator towards and away from the slit with the help of screws. This process is known as

setting of spectrometer. Make slit vertical with the help of cross wire of eye-piece.

2. Keep the prism on prism table. Its non refracting surface faces telescope and its vertex parallel to the slit of collimator. Level the telescope, prism and collimeter such that light coming through slit incident on the vertex of the prism as shown in the Fig. 6.1.

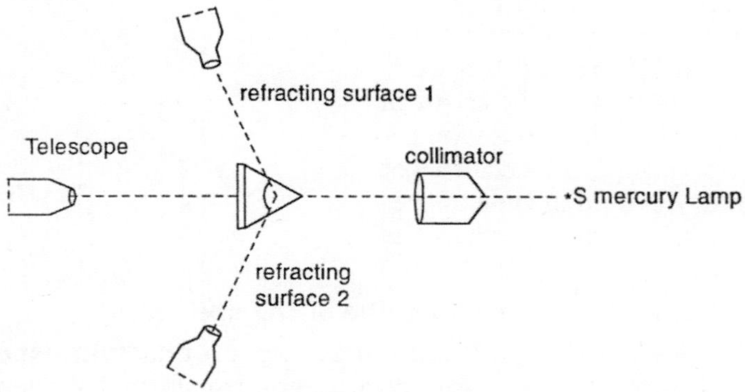

Fig. 6.1.

3. Move the telescope on either side of refracting surface and obtain the image of the slit at position T_1. Note the readings of verniers V_1 and V_2 of spectrometer. Again move telescope on other side and observe the image of the slit at position T_2. Write the observations in the observation table given as below for determination of prism angle.

OBSERVATIONS

Determination of prism angle (A)

S.No.	*Telescope on one refracting surface*		*Telescope on other refracting surface*		$V_1^0 - V_1^{10}$	$V_2^0 - V_2^{10}$
	V_1^0	V_2^0	V_1^{10}	V_2^{10}	= 2A	= 2A

Mean 2A =

A =

(B) For minimum deviation angle δm for violet, red and yellow colours

EXPERIMENTAL PROCEDURE

Follow the following steps :

1. Rotate the prism table through 135° sq that light coming from square through slit can incident on any one of the refracting surface of the prism. Incident light get deviated by the prism and dispersed through another refracting surface. Thus we obtain spectrum of white light. Move the telescope towards dispersing surface and get spectrum in the field of view of eyepiece of telescope, as shown in the Fig. 6.2.

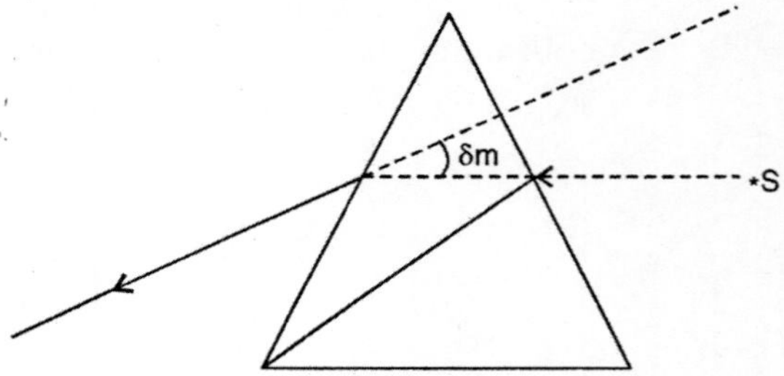

Fig. 6.2.

   focus the telescope on the violet line of the spectrum.
2. Rotate the prism table gradually and keep on observing spectrum. You, will see spectrum move in one direction. Keep on rotation of the prism table until a position is obtained from where spectrum begins to return. Fix the prism table at this very location and note down the readings of verniers V_1 and V_2 correspondingly. This is the maximum direction position for violet light. Do the same for the red light and yellow light.
3. Remove the prism from prism table and get the direct image of the slit. Note the readings of verniers V_1 and V_2 respectively. Enter the observation in a table.

OBSERVATIONS

Least count of spectrometer = ...

S.No.	V_1^0	V_2^{10}	Direct image of slit		Sm =	Sm	Mean Sm^0
			V_1^0	V_2^{10}	$V_1^0 - V'_1{}^{10}$	$V_2^0 - V'_2{}^{10}$	
Voilet							
Red							
Yellow							

CALCULATION

Using formula

$$\mu = \frac{\sin \frac{A + \delta m}{2}}{\sin \frac{A}{2}}$$

We can write

$$\mu_v = \frac{\sin\dfrac{A+(\delta m)_v}{2}}{\sin\dfrac{A}{2}}$$ Where $(\delta m)_v$ = minimum deviation for violet

$$\mu_r = \frac{\sin\dfrac{A+(\delta m)r}{2}}{\sin\dfrac{A}{2}}$$ $(\delta m)_r$ = minimum deviation for red

$$\mu_y = \frac{\sin\dfrac{A+(\delta m)y}{2}}{\sin\dfrac{A}{2}}$$ $(\delta m)_y$ = minimum deviation for yellow

Maximum probable error

$$W = \frac{\mu_v - u_r}{\mu_y - 1}$$

since $\delta = (\mu - 1)A$

hence $w = \dfrac{dv - dr}{\delta y}$

Taken log of both the sides and differentiating and taken positive value for maximum error.

$$\frac{dw}{w} \frac{d(\delta v) + d(\delta r)}{\delta_v - \delta_r} + \frac{d(\delta y)}{\delta_y}$$

$d(\delta_v) = d(\delta_r) = d(\delta_y) = 2x$ least count of spectrometer expressed in radians

$= \dfrac{2\pi}{180 \times 60}$ x lc of spectrometer expressed in minutes.

calculated dw by substituting the values of the various quantity.

RESULT

The dispersive power of the material of the prism =

SOURCES OF ERROR AND PRECAUTIONS

1. The mechanical adjustment should be correct.
2. Care should be taken in taking the readings of the two verniers.
3. The prism table and the telescope should not be rotated simultaneously. To move the one the other should be clamped.

4. Care should be taken in locating the position of minimum deviation for each spectral line.
5. Slit should always the vertical.
6. It will be better to use a thin prism made of extra dense flint glass because the theory developed is applicable in case of the thin prism. Further the dense flint glass has higher value of dispersive power which in turn, shows a larger deviation of the spectral lines.
7. The source of light should be taken which has its characteristic spectrum in the complete visible region.

VIVA-VOCE

Q.1. What is the difference between a prism and gratting spectrum ?

Ans. In a gratting spectrum violet colour is less deviated while red colour is deviated maximum. The sequence of colours in prism spectrum is reverse than that of gratting spectrum.

Q.2. What is dispersion of light ?

Ans. When a ray of light passes through a prism it splits into its constituents colours. This phenomenon is known as dispersion.

Q.3. What is the dispersive power of any material.

Ans. $\omega = (\delta v - \delta r)/\delta y$ *i.e.*, ratio of angular dispersion between the emergent violet and red rays and deviation suffered by yellow ray.

Q.4. Can you find dispersive power of prism with sodium light ?

Ans. Sodium light is a monochromatic source of light and so we can not get dispersive power.

Q.5. What are essential parts of a spectrometer ?

Ans. Callimator, prism table and telescope.

Q.6. How will you adjust the collimator ?

Ans. Callimator and the telescope are brought in a line and slit of the collimator is moved in and out by the rack and the pinion arrangement till a fine and focused image of the slit of the collimator is obtained.

Q.7. How will you level optically the prism table ?

Ans. The refracting edge of the prism should be parallel to the axis of rotation of the prism table and the slit.

Q.8. What type of eye piece is used in this spectrometer ?

Ans. Ramsden's eye-piece.

Q.9. Why do you use two verniers in this experiment ?

Ans. To remove the error in the reading that we get due to not coinciding the axis of rotation of prism table and telescope.

Q.10. What is spectrum ?

Ans. When a white light passes through prism, it splits up into its constituents of seven colours.

Q.11. How many types of spectra do you know ?

Ans. Emission spectrum and absorption spectrum.

Q.12. Do you know a relation between refractive index μ and the wavelength λ ?

Ans. Yes. There exist, the following relation

$\mu_\lambda = a + \frac{b}{\lambda^2}$, known as Cauchy's relation where a and b are constants.

Q.13. How and why does one get the spectrum of a light source when its lights is passed through a prism ?

Ans. Because of Cauchy's relation $\mu = a + \frac{b}{\lambda^2}$, longer the wavelength lower the refractive index resulting into refraction through a larger angle. In this way the various constituent wavelength of the light are arranged in order of their wavelength and one gets what is called as spectrum.

Q.14. What is angle of minimum deviation ?

Ans. It is found that as the angle of incidence of a ray i, is increased, the corresponding angle of deviation decreases initially and reaches a minimum value and these from increases. The minimum value of this angle of deviation is known as the angle of minimum deviation.

Q.15. How does a ray pass through the prism in order to suffer the minimum deviation ?

Ans. In order to suffer the minimum deviation the ray should be incident such that the angle of emergence equals the angle of incidence. In case, the prism is either isoscelens or equilaterial that the ray traverses the prism parallel to its base.

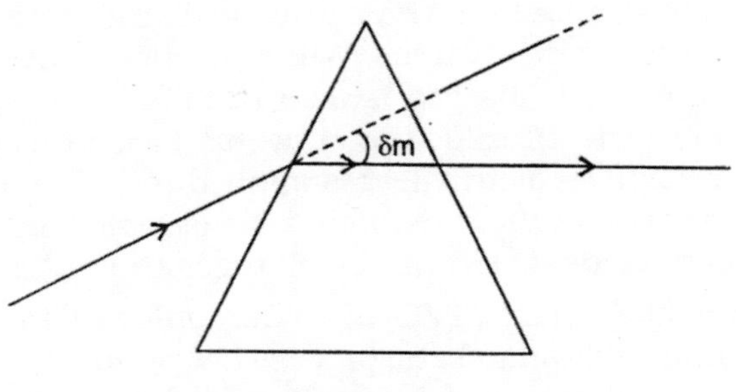

Fig. 6.3.

7

Velocity of Ultrasonic Wave in Liquid

EXPERIMENT

OBJECT

To find out the speed of the ultrasonic waves in liquids by optical diffraction method.

APPARATUS REQUIRED

Quartz crystal, sodium vapour lamp, cubical glass cell, spectrometer, radio frequency oscillator, kerosene oil.

THEORY

1. **Generation of the ultrasonic waves**

   A solid is said to be piezo-electric if elecrric charges are liberated on it when it is subjected to mechanical stress. When the sign of the stress is reversed, the polarity of the electric charges is also reversed. This phenomenon is called "Direct piezo-electric effect" ; the converse is also true, *i.e.*, the substance changes in size when electric field is applied to it, the sign if the charge being reversed on reversal of the direction of the electric field. So if an AC voltage is applied to such solids, there will be changes in its dimensions alternatively.

   Some Crystals like Quartz, PZT (Lead Zerconium Titanate), PLZT (Lead Lanthanum Zirconium Titanate) etc. process the piezo-electric property to marked degree, they do not get corroded easily and hence they are used as electro-acoustic transducers in ultrasonic experiments.

2. **Optical diffraction**

   A parallel beam of light incident on a standing wave pattern obtained by beams of ultrasonic waves in a liquid (or solid) is diffracted (Fig. 7.1) by the waves is much the same way as by a ruled grating.

   The diffraction of the *m*th order spectrum is given by

   $$\sin \alpha = m\lambda_1/2d \qquad ...(1)$$

   Where λ_L is the wavelength of the incident light and d is the grating spacing. Fig. 7.1 is a diagram of the optical system by which diffraction *i.e.*, spectrum of sodium in different orders can be observed. Light from the sources S is focused at the slit of the spectrometer to render it parallel. Then it is passed through the rectangular glass cell C, which is placed on the spectrometer table. The diffraction maxima are viewed with telescope. Before starting the experiment the spectrometer has to be adjusted. For adjustment of the spectrometer see experiment no. 2. When the liquid is at rest, a single image of the slit is seen at P. Ultrasonic waves are generated in the liquid by the excitation of the transducer T. If a monochromatic beam of light covers many wavelengths of sound, the intensity of the central image decreases, but diffracted images appear on either side of the central image, as shown in Fig. 7.1.

Explanation

The ultrasonic waves generated by the transducer T, travels in the medium (liquid), get reflected at the side face (flat glass plate) of the cell. The incident and reflected waves interfere and stationary/standing waves pattern are formed as schematically shown in Fig. 7.1. If one looks at the face of the cell from which the light emerges, the light seems to be disturbed in bright line sources with the distances *d* apart and if *d* is sufficiently small, the waves from these sources interfere like issuing from a ruled grating according to equation (1). If λ is the wavelength of US waves, it can be seen that it contains two compressions and two rarefactions, *i.e.*, it is equal to $2d$ where *d* is the corresponding grating constant.

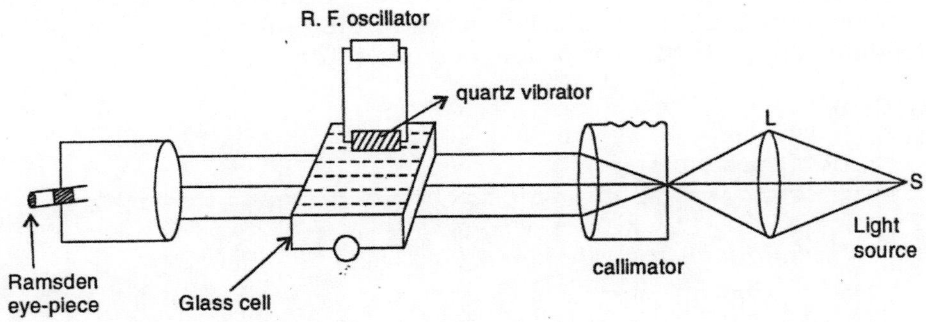

Fig. 7.1.

Working Formula

The phenomenon of diffraction of light by ultrasonic waves can be used for determining the velocity V of waves in a medium.

$$V = \nu\lambda_s \qquad ...(2)$$

Here V is the velocity of ultrasonic waves in the medium and λ_s is the wavelength of ultrasonic waves in fluid and ν (Nu) is their frequency.

For diffraction by gratting

$$(a + b) \sin \theta_n = n\,\lambda_l$$

where *n* is order of diffraction and λ_l = wavelength of light used

In the case of accoustic grating $(a + b) = \lambda_s$

$$\Rightarrow \quad \lambda_s \,.\, \sin \theta_n = n\,\lambda_l$$

or $$\lambda_s = \frac{n\,\lambda_l}{\sin\theta_n}$$

Putting the value of λ_s in eqn. (2), we get

$$\boxed{V = \frac{n\lambda_l v}{\sin\theta_n}}$$

λ_l is the wavelength of light used and θ_n is the angle of nth order diffraction. Thus velocity can be found out.

EXPERIMENTAL PROCEDURE

Switch on the sodium vapour lamp. Wait till it becomes bright. Adjust the optical alignment of spectrometer properly and get a well defined image of the slit as seen through the eyepiece of the travelling microscope. Mount the given transducer (quartz crystal) between the electrodes and adjust the set up so that the electrodes (metal plates) are exactly parallel to the bottom of the rectangular glass cell. Fill the cell with kerosene oil till the top surface of the upper electrode is well immersed. See that thee are no air bubbles below the transducer and also at the bottom surface of the lower electrodes.

Now connect the two terminals of electrodes (upper and lower) to the two terminals of the variable frequency RF oscillator. Switch on the oscillator and wait for few minutes. Keep viewing through the microscope the image of the illuminated slit and adjust slowly, the dial of the oscillator (marked from 2 to 10 MHz) until you get a very good fringe pattern as many as 6 to 7 orders on both sides of the central bright image. Take the telescope readings on the vertical scale for each other (fringe).

OBSERVATIONS

Least count of spectomerter =

Order of diffraction	*Cross-wire left to the central maxima*		*Cross-wire right to the central maxima*		$2\alpha_n = v_1 - v_1'$	$2\alpha_n = v_2 - v_2'$
	Vernier V_1	*Vernier* V_2	*Vernier* V_1'	*Vernier* V_2'		
4	...	...	...	...		
3	...	...	...	...		
2	...	...	...	...		
1	...	...	...	...		

$$\theta_n \text{ for given order} = \frac{2\alpha_n + 2\alpha_n'}{4}.$$

VIVA-VOCE

Q.1. What is ultrasonic wave ? What is its nature ?

Ans. Ultrasonic waves are those sound waves which have frequency greater than 20 kHz. These are longitudinal waves.

Q.2. What is piezo-electric effect ?

Ans. 'Piezo' is a Greek word having meaning 'Pressure'. It has been observed in a class of crystal like quartz, that if mechanical pressure is applied along a characteristic axis called mechanical axis then e.m.f. is automatically generated along a perpendicular axis called electrical axis. This phenomena is called Piezo-electric effect. Similarly if e.m.f. is applied along electrical axis then mechanical pressure (compression or extension) is generated along mechanical axis. It is called inverse piezo-electric effect.

Q.3. Wheather you are using piezo-electric or inverse piezo-electric effect in this experiment.

Ans. Inverse piezo-electric effect.

Q.4. What is acoustic gratting ?

Ans. When ultrasonic waves are generated by inverse piezo-electric effect then they propagate through the liquid medium (kerosene oil) and reach to the walls of container. At the walls they get reflected and are superposed to the initial ultrasonic waves and hence stationary waves are generated across the medium. As a result medium possesses high density region (compression) and low density region (rarefaction) which will be occuring alternatively. If monochromatic light is allowed to incident on such a medium it gets diffracted and one can have diffraction pattern. This type of arrangement in the liquid medium which arises because of ultrasonic wave is called acoqustic gratting.

Q.5. What will happen if we use white light instead of monochromatic light in this experiment ?

Ans. We can not obtain an explicit diffraction pattern.

Q.6. What is velocity of ultrasonic wave in vaccum, in air and in liquid ?

Ans. Zero metre/sec in vaccum, 332 m/s in air and 1447 m/s in water.

8

Carey Forster's Bridge

EXPERIMENT

OBJECT

To determine

1. resistance per unit length of the wire of Carey Foster's bridge,

2. specific resistance of a wire.

APPARATUS REQUIRED

Carey Foster's bridge, decimal ohm box, two standard resistance coils of 1 or 2 ohm each, screw gauge, metre scale, Lechlanhe cell, galvanometer.

FORMULA USED

1. The resistance per unit length of the wire of Carey Foster's bridge is given by X

$$\sigma = \frac{X}{l_2 - l_1} \quad \text{...(1)}$$

Where X = Resistance l_1 and l_2 are the lengths at which the balances point is obtained.

2. The specific resistance of wire is given by

$$\rho = \pi r^2 R/l \quad \text{...(2)}$$

Where R = Resistance of the wire at room temperature, l and r are respectively the length and radius of the wire.

Basic Principles Involved

Carey Foster's Bridge. This is a modified form of meter bridge and is free from the error due to the end resistances. Let α and β be the end resistances and let the null point be obtained at a distance l_1 from the left hand end of the wire. Then in view of the conditions of wheat stone's bridge, we get

$$\frac{P}{Q} = \frac{X + \alpha + \sigma l_1}{Y + \beta + \sigma(100 - l_1)}$$

here σ represents the resistance per unit length of the wire. Let the balance point on interchanging the two resistances X and Y be obtained at distance l_2 then

$$\frac{P}{Q} = \frac{X + \alpha + \sigma l_1}{Y + \beta + \sigma(100 - l_2)}$$

on equating the right hand sides of equations and simplifying we have

$$X - Y = \alpha(l_2 - l_1)$$

This is a basic equation. It is employed to know the value of an unknown resistance X by determining the length l_1 and l_2 corresponding to a known resistance Y ; provided σ is known. In order to know σ one may determine the length l_1 and l_2 corresponding to known resistance X and Y and use equation it is customary to use a thick copper plate in place of Y. In this case Y = 0 and consequently equation yields.

EXPERIMENTAL PROCEDURES

1. **Determination of σ.** make connections as shown in Fig. 8.1 use P = Q say 1 or 2 ohm each. In place of X and Y put a decimal ohm and thick copper strip respectively. Give some suitable value to X and find the balance point. Let this length be l_1 as measured from the left end. Take several observation for l_1 by changing the value of X. Now, interchange the decimal ohm box and the copper strip and again find the balance point for all the value of X used earlier. Measure the distance l_2 of the corresponding balance point from the left end. Calculate σ using equation for all the observation and finally find its average value.

   An alternate method can also be adopted in evaluating the value of σ. Take P = Q = 2 ohm and Y = 1 ohm and obtain the balance point l_1 keeping Y fixed, take several observaticn by changing the value of X interchange X and Y and find out l_2 for all earlier values of X by using the relation viz $A = Y + \sigma(l_2 - l_1)$, σ can be evaluated.

2. **Determination of specific resistance of a wire.** Connection are to be made as in Fig. 8.1. Connect the wire under investigation in place of X and a decimal ohm box in place of Y. Take P = Q = 2 or 1 ohm each give some suitable value of Y and find out the balance point. Let this length be l_1 by changing Y take at least five observation. Interchange the position of decimal ohm box and the wire under investigation. Find out the balance point and not the length l_2 for all pervious value of Y used in determining l_1. Calculate the resistance of the wire l R_1 using equation l viz., $R = Y + \sigma(l_2 - l_1)$. Determine length l and radius r of the wire with the help of a metre scale and a screw gauge respectively. The specific resistance may be calculated with the help of the equation.

$$\rho = R\pi r^2/l$$

Q.4. What do you mean by specific resistances of a wire and what is its unit ?

Ans. The specific resistance of a wire is the resistance per unit length per unit cross-section. Its unit is ohm-cm and ohm-metre in C.G.S. and M.K.S. system respectively.

Q.5. What is specific resistance ?

Ans. Specific resistance of a material of wire is the resistance of one centimetre length of the wire whose cross-sectional area is the one square centimetre.

Q.6. What is resistance ?

Ans. Resistance $R = \frac{V}{I}$ where V is the potential difference applied across the ends of the wire and an electric current of 1 amp. flows through it.

Q.7. What is the effect of temperature on resistance ?

Ans. Resistance increses with increase of temperature of the material of wire.

Q.8. In what materials, the resistance decreases with increasing temperature ?

Ans. Resistance of semiconductors decreases with increase of temperature.

Q.9. If the radius of the wire is doubled, will the specific resistance change ?

Ans. Specific resistance is independent of length or radius of the wire and depends only on material of the wire.

Q.10. If not explain it.

Ans. No, because specific resistance is a constant of proportionalty and is property of the wire.

Q.11. How are the resistance coils placed inside a resistance box ?

Ans. Insulated wires of suitable lengths of maganin are doubled on itself and then it is wound on a non-conducting material. The free ends of the wire are soldered to the piece of the brass at the top.

Q.12. Why is the resistance wire doubled before winding over the bobbin ?

Ans. To avoid the inductive effects. Direction of current in one wire is opposite to the other so the total flux is always zero.

Q.13. What is the principle of Carey Foster's bridge ?

Ans. Position of null point changes on the bridge wire when the resistances connected in the outer gaps are interchanged. The difference in the value of resistance is equal to the resistance of the differences of bridge wire *i.e.*, $P(l_2 - l_1)$.

Q.14. What is the principle of Wheatstone's bridge ?

Ans. When the resistance P, Q, R and S is the four arms of Wheatstone bridge are such that there is no deflection in the galvanometer we have

$$\frac{P}{Q} = \frac{R}{S}.$$

Q.15. When is the C.F. bridge most sensitive ?

Ans. Carey Foster's bridge is most sensitive when P, Q, R and S are equal or of the same order.

Q.16. What is the maximum difference in the two resistances that can be measured with it ?

Ans. It is equal to the total resistance of the Carey Foster's bridge wire.

Q.17. What is the minimum difference in resistance that can be measured with it ?

Ans. The resistance of one milimetre length of the bridge wire can be measured.

Q.18. Why the resistance of inner ratio arms should be equal ?

Ans. Sensitivity and accuracy of the Carey Foster's bridge is increased.

Q.19. What are the end resistances ?

Ans. The resistance of the copper strips fitted at the end of the wire and that of solder are not taken into account. These are called end resistance.

Q.20. What is the advantage of interchanging the position of resistance box and the unknown resistance in the outer gaps ?

Ans. See Ans. 9.

9

Steward Tangent Galvanometer

EXPERIMENT

OBJECT

To determine the variation of magnetic field along the axis of current carrying coil and then to estimate the radius of the coil.

APPARATUS REQUIRED

Tangent galvanometer of the Steward and Gee type, battery, rheostat, commutator, plug key.

FORMULA USED

$$F = \frac{2\pi n r^2 i}{10(x^2 + r^2)^{3/2}}$$

F = field along the axis of the coil

n = number of turns in the coil

r = radius of the coil

i = current in ampere flowing in the coil

x = distance of the point from the center of the coil

$$F = H \tan \theta$$

(if F is perpendicular to H, the horizontal components of the earth's magnetic field).

EXPERIMENTAL PROCEDURE

Set the coil in the magnetic meridian and place compass box on the sliding bench so that its magnetic needle is at the center of the coil. Rotate the compass box toll the pointer ends read 0.0 on the circular scale. Now with the help of rheostat (Rh) adjust the current such that the deflection of nearly 70° to 75° is produced in the compass needle. Read both the ends of the pointer. Reversing the direction of the current again, read the both ends of the pointer. Mean deflection is the mean of the four readings at $x = 0$. Now, shift the position of the compass needle on either side of the axis of the coil and note down the mean deflection. Plot a graph between distance from the centre of the coil (x) along X-axis and tan θ along Y-axis and mark the points of the inflexion on the curve. The distance between the two points will be the radius of the coil.

CALCULATION

Graph plotted between x and tan θ. The distance between the points of inflexion MN is the radius of the coil.

OBSERVATIONS

S.No.		*Deflection on west arm*						*Deflection on east arm*					
	Distance of needle from 'x'	*Current in one direction* θ_1	θ_2	*Current reversed* θ_3	θ_4	*Mean* $(\theta_1+\theta_2+\theta_3+\theta_4)/4$	*tan* θ	*Current in one direction* θ_1	θ_2	*Current reversed* θ_3	θ_4	*Mean* $(\theta_1+\theta_2+\theta_3+\theta_4)/4$	*tan* θ

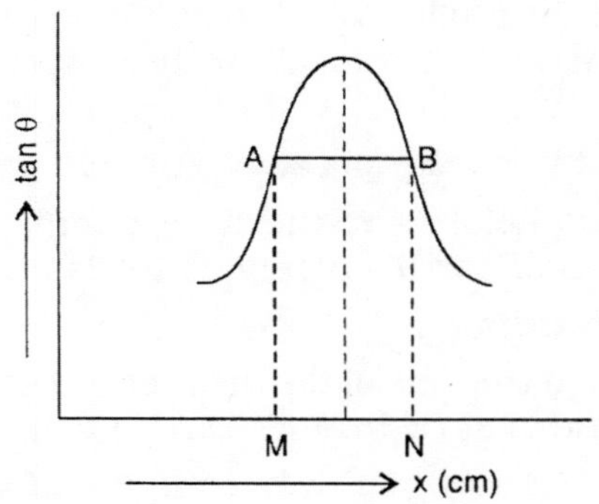

Fig. 9.1.

VIVA-VOCE

Q.1. How does the field vary along the axis of the coil ?

Ans. Magnetic field F at a distance x from the centre of the coil of radius r is given as

$$F = \frac{2\pi n i r^2}{10(r^2 + x^2)^{3/2}}$$

where n is the number of turns and i is the current flowing through the coil.

Q.2. What is the magnitude of the field at the centre ?

Ans. For F at centre $x = 0$

$$F = \frac{2\pi ni}{10r}.$$

Q.3. How can you increase the region of uniform field.

Ans. By using two identical vertical flat circular coils placed co-axially at a distance equal to the radius of the either coil.

Q.4. Whether the field is uniform at the centre or not ?

Ans. Field is maximum at the centre and decrease on either side.

Q.5. What precautions should be taken in this experiment.

Ans. (*i*) Plane of coil should be set in magnetic meridian.

(*ii*) Current flowing in the circuit should be constant and the value of current should be such that there is deflection of 70°.

(*iii*) For every observation, position of both the pointers of the compass needle should be noted.

Q.6. What is the strength of magnetic field at any point on the axis of circular coil ?

Ans. See Ans. 1.

Q.7. What is the strength of magnetic field at the centre or the coil ?

Ans. See Ans. 2.

Q.8. What is tangent law ?

Ans. $F = H \tan \theta$ where θ is the angle made by magnetic needle with horizontal component of earth's magnetic H.F. and H are two mutually perpendicular uniform magnetic fields.

Q.9. Why is it necessary to set the plane of the coil in magnetic meridian ?

Ans. To make the magnetic field F produced by circular coil at right angles to the horizontal components of earth's magnetic fields.

Q.10. What is magnetic meridian ?

Ans. A plane passing through the axis of the magnetic needle freely suspended through its centre of gravity and is at rest under earth's field is called magnetic meridian.

Q.11. What is the practical utility of the knowledge of variation of magnetic field along the axis of the coil ?

Ans. It shows the existence of two points of inflexion where the rate of change of magnetic field with distance is constant. We made use of this property the construction of galvanometer.

Q.12. What is Helmholtz galvanometer ?

Ans. In Helmholtz galvanometer there are two identical vertical flat circular coils placed co-axially at a distance equal to the radius of the either coil. Midway between the two coils magnetic field is very nearly constant where the compass box is placed.

Q.13. What is reduction factor ?

Ans. We get the current through the coil by multiplying reduction factor with the angle of deflection. Reduction factor $K = 2rH/\mu_0 N$ where r is the radius of the coil,

H is horizontal component of the earth, N is number of turns in the coil, μ_0 is the permeability of free space.

Q.14. What is point of inflexion ?

Ans. It is the point on the x – tan θ curve where the curve changes its direction of curvature.

Q.15. What is distance between the two points of inflexion ?

Ans. It is equal to the radius of the circular coil.

Q.16. Why is a mirror provided below the pointer in the compass box ?

Ans. It is used for measuring the exact deflection of the magnetic needle by removing the parallex between the pointer and its image.

Q.17. How is the error in the measurement of deflection is eliminated if the place of coil is not exactly in magnetic meridian ?

Ans. By passing current in one direction and then in reverse direction and taking the mean of these four readings, we get deflection free from error.

Q.18. What are the advantage of a Helmholtz galvanometer over a tangent galvanometer ?

Ans. (*i*) In the case of Helmholtz galvanometer field is almost constant between the two coils while in tangent galvanometer it is constant at the centre.

(*ii*) Sensitivity of Helmholtz galvanometer is more.

(*iii*) For the same current Helmholtz galvanometer produces more field and hence greater deflection.

10

Callendars and Griffith's Bridge

EXPERIMENT

OBJECT

To determine the temperature coefficient of resistance for platinum using a platinum resistance thermometer, and Callendars and Griffith's bridge.

APPARATUS REQUIRED

Callender and Griffith's bridge, platinum resistance thermometer, galvanometer, hot water bath, cell plug key and connection wires.

FORMULA USED

Temperature coefficient of resistance, α is given by the formula

$$\alpha = \frac{R_2 - R_1}{R_1 t_2 - R_2 t_1} \text{ per } {}^0C$$

Where R_1 = Resistance of platinum wire at t_1 °C

R_2 = Resistance of platinum wire at t_2 °C

Using a Callender and Gnffth's bridge, the resistance(R) is given by the formula

$$R = r \pm 2\rho x$$

where P = Resistance per unit length of the C and G bridge potentiometer wire.

r = Resistance introduced in resistance box.

X = Distance of null point from middle point (electrical zero).

The proper sign is used according to the balance point lies at right or left of the electric zero.

THEORY

1. Platinum resistance thermometer. The resistance of an electric conductor is a function of temperature and forms the basis of resistance thermometry. Many of conductors do not offer always the same resistance at a given temperature but the resistance of a platinum wire free from impurities is found to be a unique function of temperature. The platinum resistance thermometer is shown in Fig. 10.1. Pure platinum wire free form impurities (carbon, silicon, tin and others) double on itself to avoid induction effect, is wound on a thin plate of mica. The ends of the platinum wire are soldered to two thick leads of copper or platinum whose other ends are connected to two terminals PP on the top or porcelain tube. Parallel to the thick leads there are two extra identical leads (same material and size) whose other ends are connected to the terminals CC. This extra identical leads is called compensating leads. The function of compensating leads is to eliminate the resistance of the connecting leads to the platinum wires at all temperature. A number of mica discs are placed in the tube to insulate the leads from each other. This also prevents convection current in the air of the tube.

2. Callender and Griffith's bridge. For the measurement of the resistance of platinum wire at different temperature, Callender and Griffith devised a modified form of Wheatstone bridge known as Callender and Griffith's bridge, this is shown in Fig. 10.1(*a*) and (*b*).

In the circuit, the ratio arms P and Q are made equal. T platinum leads are connected in third arm S., while the fourth R contains coils of different known resistance (marked) 1, 2 ...... 10 unit etc. (each unit is equivalent to the stretched wire) along with the compensating leads of the thermometer. Between the arm R and S a uniform resistance wire MN resistance ρ ohm per cm is connected. A sliding contact *d* moves on the wire MN. The battery B and galvanometer G are also connected as shown in the figure.

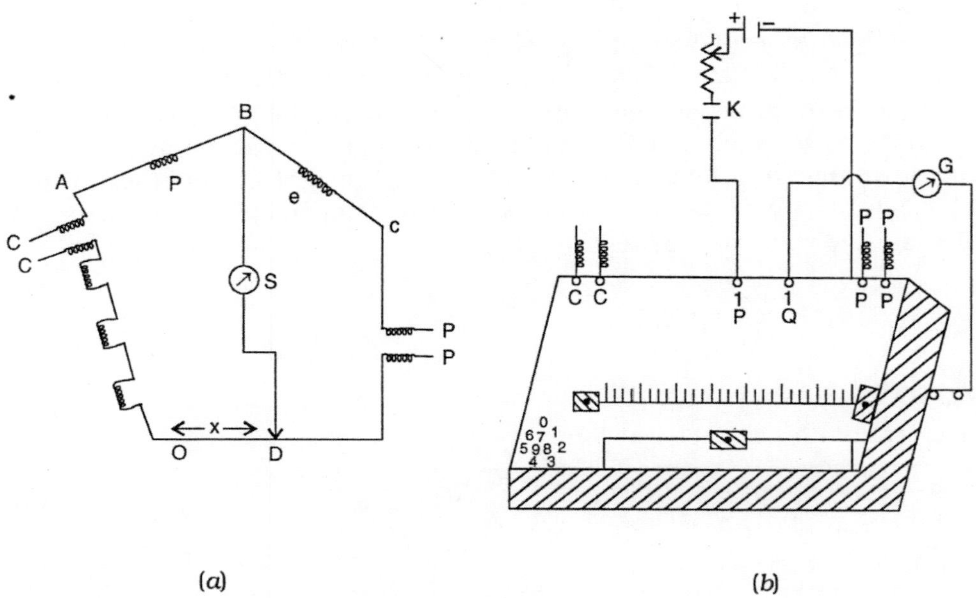

(*a*) (*b*)

Fig. 10.1.

EXPERIMENTAL PROCEDURE

1. **Determination of electrical zero.** Complete the connections as shown in Fig. 10.1(*b*), short circuit the gaps PP and CC *i.e.*, connect the terminal by a thick copper wire. Make the resistance R zero by moving the key of variable resistance. Now make the cell circuit on by inserting the plug key. Deter-

mine the position of null point by moving the jockey on the bridge wire. The null point gives the position of electrical zero. This should be noted down.

2. **Determination of resistance per unit length (ρ) of the bridge wire.** To determine the value of ρ, the gap (PP) and (CC) are short circuited. The value from R is adjusted to a small value r to obtain the balance. Let the balance be on the left side of O at a distance x', from the position of electrical zero.

$$0 = r - 2\rho x'$$

$$\rho = \frac{r}{2x'}$$

3. **Determination of R_1 and R_2 at steam temperature and room temperature respectively.** Connect the compensating leads and platinum leads of platinum resistance thermometer to the Callendards and Griffith's bridge at CC and PP respectively. Other connections are made according to the circuit diagram shown in Fig. 10.1(*b*). Place the bulb of the thermometer in a tumbler of ordinary water bath at room temperature and wait for some time to ensure that it has acquired the temperature of water. Note the temperature of water with the help of mercury thermometer. By introducing a suitable resistance, in the circuit of compensating leads with the help of moving coil resistance obtain the balance. The length of the bridge wire is noted. Using the following formula the resistance of platinum wire R_1 at room temperature is calculated.

$$R_1 = R' \pm 2\rho x$$

Where R′ = Resistance introduced in the variable resistance, R.

X = Length on bridge wire from electrical zero to get the null point.

Note the platinum resistance thermometer is placed in a hypsometer or in a hot water bath and wait for such time that it acquires the constant temperature of bath. Note this temperature also and determine again the resistance of the platinum wire as described above in the step 2. Let the resistance at this temperature be denoted by R_2.

Calculate the value of α using the relation.

$$\alpha = \frac{R_2 - R_1}{R_1T_2 - R_2T_1}.$$

OBSERVATIONS

1. Determination of electrical zero

S.No.	*Position of balance point (cms) when both PP and CC are shorted*	*Electrical zero (cms)*	*Mean electrical zero in (cms)*
1.			
2.			
3.			
4.			

2. Determination of ρ

S.No.	Resistance introduced in the moving coil resistance, (R)	Distance of null point on the bridge wire (cms)	Distance between electrical zero and null point (cms)	$P = \frac{R}{2x'}$	Mean P ohm/cm
1.					
2.					
3.					
4.					

3. Table of the determination of R_1 and R_2

S.No.	Temperature of bath	Resistance used in coil	Reading	Shift balance from zero x (cms)	Resistance of platinum wire in the thermometer (ohm)	Mean (ohm)
1.	Room temp.	2.0			R_1 =	R_1 =
2.	= °C	3.0			R_2 =	
3.						
4.						
5.						
1.	Temperature	2.0			R_2 =	R_2 =
2.	of boiling	3.0			R_2 =	
3.	water = °C					
4.						
5.						

CALCULATION

$$\text{Temperature coefficient } \alpha = \frac{R_2 - R_1}{R_1T_2 - R_2T_1}.$$

RESULT

The temperature coefficient of resistance for platinum = ... °C.

Standard result. The standard value of temperature coefficient of resistance for platinum = 37×10^{-4} °C.

SOURCES OF ERROR AND PRECAUTIONS

1. The end of the connection wires as well as surfaces of copper strips should be well cleaned.
2. The balance point for the measurement of R_1 and R_2 should be determined only when the temperature acquired by the platinum resistance thermometer are steady.
3. Jockey should be pressed on the brdige wire so as to make proper contact.
4. The jockey should never be kept pressed while it is being moved on the wire.
5. Protect the galvanometer with a shunt while approximate null point is in the process of obtaining.

11

(e/m) of an Electron by Thomson Method

EXPERIMENT

OBJECT

To determine the value of specific charge (e/m) of an electron by Thomson method.

APPARATUS REQUIRED

Cathode ray tube (CRT) with its power supply unit. Pair of bar magnets, mangetometer, compass box, a wooden stand to place the bar magnets on its two arms with CRT in middle.

THEORY

Let OX be the undeviated path of the electron of mass '*m*', charge '*e*', velocity '*v*', if an uniform magnetic field be applied parallel be OZ at right angles to the plane of the paper the electrons will be deflected in the same manner as the current carrying conductor in the magnetic field. Therefore, the force on an electron causing upward deflection along the circular path of radius *r*, is given by

$$\frac{Mv^2}{R} = Hev \qquad \text{...(1)}$$

If both electric and magnetic field are applied simultaneously, they may be so adjusted that the effect of each of the fields nullify each other. *i.e.*,

$$Hev - Ee = 0 \qquad \text{or,} \qquad v = E/H \qquad \text{...(2)}$$

Substituting the value of v in equation (1) we have

$$\frac{e}{m} = \frac{E}{rH^2} \qquad \text{...(3)}$$

to find the value of r (*i.e.*, radius of the circular path in the magnetic field) the deflection 'y' of the lumonious spot on the screen is to be noted if l be the distance of the center of the screen from the center of the magnetic field and l be length of the deflection plate, r is given by,

$$r = \frac{Ll}{y} \qquad ...(4)$$

Substituting in (3), we get,

$$\frac{e}{m} = \frac{yE}{LlH^2} \qquad ...(5)$$

the electric field is given by, E = V/*d*, where 'V' is voltage of the plate and '*d*' is the distance between the plate. Substituting in (5), we have,

$$\frac{e}{m} = \frac{yV}{Lld B^2}$$

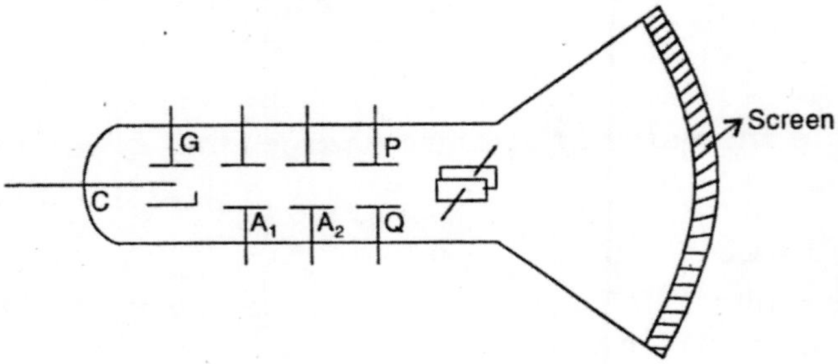

Fig. 11.1. Cathode Ray Tube (CRT).

C → Cathode

G → Control grid

P, Q → Deflecting plates

A_1 → First anode (Focusing anode)

A_2 → Second anode (Accelerating anode).

FORMULA USED

$$\frac{e}{m} = \frac{V \times 10^7 \times y}{LlH^2 d} \text{ e.m.u./gm}$$

Where various parameters are,

l = length of the deflection plate

L = distance of screen from the edges of the plate

Y = total deflection of spot on the screen

H = intensity of applied field

D = separation between the plates P and Q.

EXPERIMENTAL PROCEDURE

(A) First part

1. Place CRT in wooden stand such that CRT faces towards the magnetic north and south while wooden stand towards east and west.

2. Switch on the power supply and focus the spot on the screen.
3. Read the initial reading of the spot on the scale attached with screen, say it is – 0.3 (it can be zero also. So take the reading whatever it is), now give the deflection in upward direction with the help of deflecting voltage (knob marked 'acceleration' on power supply) such that the final reading is + 0.7 so that the deflection 'y' on the screen of the spot is (0.3 + .07) = 1.0.
4. Note this applied deflecting voltage (for deflection 0.7 from this voltmeter of power supply, this is 'V'.
5. Now place bar magnets deflecting symmetrically on each side such that their opposite poles face each other and their common axis is perpendicular to the axis of common axis is perpendicular to the axis of CRT. Adjust distance and polarity of the magnets such that the spot traces back to its initial position (which was – 0.3 cm). Note the distance r_1 and r_2 of the poles of the magnets nearer to the CRT on the scale fixed in the wooden stand.
6. Remove the magnets and reverse the polarity of 'V' applied to the deflecting plates PQ. Now initial deflecting reading is 0.3, and final be 0.7.
7. Place the magnets again on the stand and adjust them to trace the spot back to the initial reading (+ 0.3 cm). Note the distance r_1 and r_2 from the magnets closer to CRT.

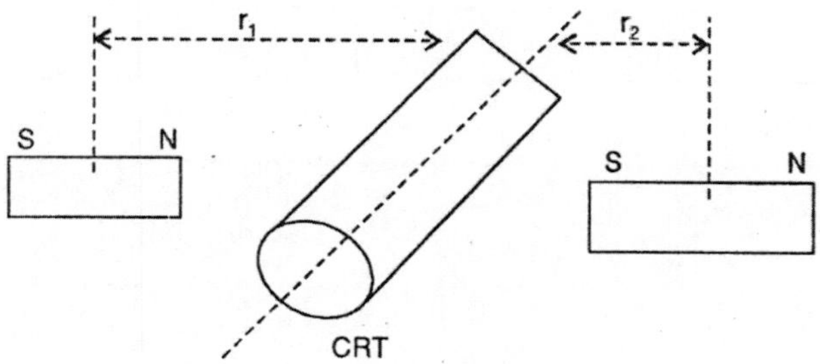

Fig. 11.2.

(B) Second part : to measure the magnetic field H

1. Remove CRT and magnets from stand and place a magnetometer compass box such that its centre lies on the common axis of the magnets. Adjust the pointer of the compass box to read 0.0.
2. Now place magnets exactly as they were in previous arrangement corresponding to distance r_1 and r_2 respectively. Read both ends of the pointer, say they are θ_1 and θ_2.
3. After this place the poles at r_1' and r_2' exactly as placed in point (*vi*) of the first part. Note the deflection of the both ends of the pointer, say they are θ_3 and θ_4.
4. Mean deflection of the deflecting voltage 'V' is calculated by taking mean of θ_1, θ_2, θ_3 and θ_4, *i.e.*, $\theta = (\theta_1 + \theta_2 + \theta_3 + \theta_4)/4$. So that magnetic field is H = Hc tan θ, where Hc is the earth's horizontal component of the field at the place of exponents.

   For various sets, vary deflecting voltage V (therefore different values of y and repeat the whole procedure as laid down in parts (A) and (B).

OBSERVATIONS

1. Constant of the tube
   (a) Separation between the plates $d = 0.4$ cms
   (b) Length of horizontal pair of plates $l = 2.0$ cm
   (c) Distance of screen from the edges of the plate P, Q L = 16 cm
   (d) Horizontal component of the earth's magnetic field at the place :
   Hc = 0.336 oersted

2. Determination of deflection y for different values of voltage V

S.No.	*Applied voltage V*	*Initial position of the spot (cm)*	*Final position of the spot after deflection cm*	*Deflection on y cm*	*R_1 and pole cm*	*R_2 and pole cm*	*R_1' and pole cm*	*R_2' and pole cm*
1.	V_1							
2.	V_2							
3.	V_3							
4.	V_4							
5.	V_5							

Determiantion of H

S.No.	*Voltage V*	*Reading of two ends of pointer of compass box*				*Mean deflection $(\theta_1 + \theta_2 + \theta_3 + \theta_4)/4$*	*Field Hc tan θ*
		θ_1	θ_2	θ_3	θ_4	$\frac{\theta_1+\theta_2+\theta_3+\theta_4}{4}$	*Hc tan θ*
1.							
2.							
3.							
4.							
5.							

CALCULATION

$$\frac{e}{m} = \frac{V \times 107 \times y}{LlH^2d} \text{ e.m.u./gm}$$

RESULT

e/m for an electron = ... e.m.u./gm = ... columb/kg.

Standard result. Standard value of $e/m = 1.76 \times 10^7$ e.m.u./gm

$$= 1.76 \times 10^{11} \text{ coulomb/kg}$$

SOURCES OF ERROR AND PRECAUTIONS

1. The CRT should be handled carefully.
2. CRT should point magnetic N-S whereas wooden stand must lie perpendicular to each other but in same horizontal plane.
3. Axis of the magnets and axis of the tube must lie perpendicular to each other but in the same horizontal plane.
4. The electric field between the plates can not be uniform because separation between them is small.

12

Frequency of A.C. Mains by Sonometer

EXPERIMENT

OBJECT

To determine the frequency of A.C. mains with the help of sonometer.

APPARATUS REQUIRED

Sonometer, A.C. square, weight-box, pan, horse shoe magnet, step down transformer, screwgauge.

FORMULA USED

Frequency of A.C. mains at resonance is given by

$$n = \frac{1}{2l}\sqrt{\frac{T}{M}} \quad \Rightarrow \quad \boxed{n = \frac{1}{2l}\sqrt{\frac{T}{M}}}$$

where l = effective length of string

$T = mg$ tension applied on the wire

M = mass per unit length of the wire of the sonometer

$= \pi r^2 d$

r = radius of the wire

d = density of the wire.

EXPERIMENTAL PROCEDURE

1. Connect two ends of the wire of sonometer across the secondary of the transformer. Place the magnet at the middle of the wire such that wire lies between two poles of the magnet. Place a mass m_1 in the pan. Adjust the distance between two knife edges such that the wire vibrate with maximum amplitude. Note the separation.
2. Again place mass m_2 on the pan and repeat the process as described above. Similarly take observations for masser m_3, m_4 ..... and so on.
3. Find out radius of the wire with the help of a screwgauge. Enter these readings in the observation table as given below.

OBSERVATIONS

1. For measurement of 'l'

Least count of the scale of sonometer = ...

S.No.	Mass suspended (gm)	Position of Ist bridge x_1 (cm)	Position of second bridge x_2 (cm)	$L = x_1 - x_2$ cm	$T = mg$ dyne	$\frac{\sqrt{T}}{l}$
1.						
2.						
3.						
4.						

2. For measurement of r

Least count of screwgauge = ...

Mean = $\frac{\sqrt{T}}{l}$

Along one direction		$r = \frac{a+b}{2}$ cm	Along perpendicular direction		
Main scale reading (cm) (a)	Vernier scale reading (cm) (b)		m.s. reading (cm) a'	v.s. reading (cm) b'	$r = \frac{a'+b'}{2}$ cm

Mean r = ... cm

CALCULATION

$$n = \frac{1}{2l}\sqrt{\frac{T}{M}}$$

$$= \frac{1}{2\sqrt{M}}\left(\frac{\sqrt{T}}{l}\right)$$

$$n = \frac{1}{2\sqrt{\pi r^2 d}}\,\frac{\sqrt{T}}{l}$$

Putting values of all terms, n can be calculated.

Maximum problem error

$$n = \frac{1}{2l}\sqrt{\frac{T}{m}}$$

$$n = \frac{1}{2l}\sqrt{\frac{T}{\pi r^2 d}}$$

taking logarithm of both the sides

$$\log n = \log 2 - \log l - \frac{1}{2}\log T - \log \pi - \log r \log d$$

$$= -\log 2 - \log l - \frac{1}{2}\log m - \frac{1}{2}\log g - \log \pi - \log r - \log d.$$

Differentiating

$$\frac{dn}{n} = -\frac{dl}{l} - \frac{ds}{r}$$

for maximum error, taking + ve sign

$$\frac{dn}{n} = \frac{dl}{l} + \frac{dr'}{r'}$$

Since position of x_1 and x_2 are being measured hence there would be twice dl to account maximum error and similar for s, hence

$$\frac{dn}{n} = \frac{2dl}{l} + \frac{2dr}{r}$$

$$dn = \left(\frac{2dl}{l} + \frac{2dr}{r}\right)n$$

where dl = l.c. of metrescale of sonometer

dr = l.c. of screwgauge.

SOURCES OF ERROR AND PRECAUTIONS

1. Place the magnet symmetrically about the wire such that its pole pieces lie in the plane of the wire.
2. There should be no kink in the wire.
3. The wire should pass over the bridge smoothly.
4. For getting the correct value of l, the length of a single loop it is important to form well defined loops either by changing the length of the wire of by changing the mass placed on the pan.

VIVA-VOCE

Q.1. What is the principle of this experiment ?

Ans. The basic principles of the experiment are forced oscillation and stationary waves.

Q.2. What do you understand by forced oscillations ?

Ans. An oscillating system because of damping forces ceases to vibrate after some time. In order that system oscillates continuously and uniformly this loss of energy has to be compensated by external agencies. The forced oscillation is such a phenomenon where we apply an external source to the oscillating system. In the beginning the system oscillates with its own natural frequency but soon it starts oscillating with the frequency of external source.

Q.3. What is the purpose of using a step down transformer ?

Ans. In step down transformer the output voltage is considerably reduced depending upon the number of turns in the secondary of the transformer. In the present experiment the A.C. main supply which is at 220 volts is reduced to 6 to 8 volts by a step down transformer. This is necessary otherwise there will be a large heating effect in the wire and also because it reduces the risk of shock while doing the experiment.

Q.4. What do you understand by phenomenon of resonance ?

Ans. When we apply an external force on an oscillating system, it vibrates in the steady state with the frequency of the external source. The amplitude of vibration depends on the frequency. There occurs a frequency at which the amplitude of the vibration is maximum. This phenomenon is called resonance and the corresponding frequency is called resonant frequency. It happens at the natural frequency of the oscillating system.

Q.5. Explain how are stationary waves formed ?

Ans. The wave which propagates from one bridge gets reflected at the other bridge. These two waves, which propagate in opposite direction and are of equal amplitude, superpose and stationary waves are thus formed.

Basic principles invovled. The A.C. main is connected to the primary of step down transformer whose secondary is connected to the ends of sonometer wire, the wire is fixed at one end and carries a hange with weight at the other end after passing over two bridges. In this way a known and constant tension has been applied to the wire. A U shaped permanent magnet is placed in between the sonometer wire such that (*i*) its poles lie symmetrical to the wire. (*ii*) the magnetic field acts in the horizontal plane containing the wire and (*iii*) it is perpendicular to the wire. On switching on the mains a current flows in the wire along its length according to Fleming's left hand rule the wire will move in a direction perpendicular to the direction of current and applied magnetic field. In view of the Fig. 12.1 the wire will move in the vertical plane. As the direction of current alternates, the wire reverses its motion and moves up and down. The wire has a number of ferquencies depending upon (*i*) its length and (*ii*) the tension on the wire ; and due to this, the resonance can occur at any of the frequencies. By changing the separation of two bridges we can change the length of the wire in between. These bridges serve as fixed ends of the wire and the wave travelling from one bridge suffers reflection at the other. When the frequency of the wire becomes equal to that of the external force *i.e.*, of the A.C. supply, there arises the phenomenon of resonance and stationary waves are formed between the bridges see Fig. 12.1. Let be the length of the wire for a single loop, T be the tension applied to the string and m be the mass of the wire per unit length, then the frequency of A.C. main would be given by

$$n = \frac{l}{2l}\sqrt{\frac{T}{M}}$$

Knowing the various parameters of the above equation, the frequency of A.C. main can be determined.

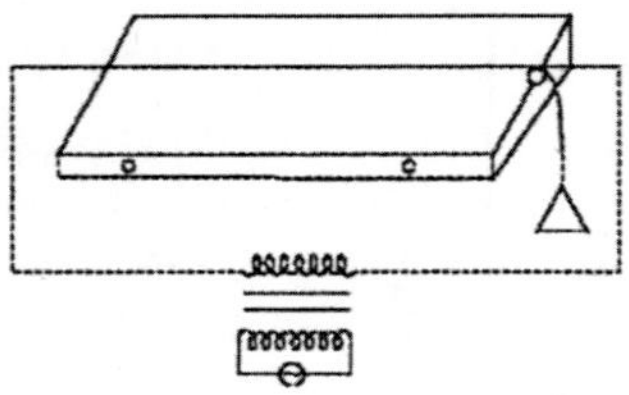

Fig. 12.1. A.C. Mains.

13

To Study Hall Effect

EXPERIMENT

OBJECT

To determine hall coefficient, carrier density and mobility of a given semi conductor material.

APPARATUS REQUIRED

Search coil, calibrated flux meter, millivoltmeter, battery, rectangular slate of semi-conductor crystal, ammeter, and keys.

FORMULA USED

$$\text{Hall coefficient} = R_H = \frac{V_H \cdot d \times 10^4}{I_x B_\mu}$$

Where,

V_H is the hall voltage measured in volts

I_x is measured in amperes

D is thickness of the crystal along within magnetic field and measured in metres

$B_\mu = \mu B$ in gauss.

Carrier density (no. of charge carrier per unit volume)

$$N = \frac{1}{R_H \dot{e}}$$

Where '$\dot{e}$' is the electronic charge

$$\dot{e} = 1.6 \times 10^{-19}$$

Mobility of carriers

$$\mu = \frac{\varphi}{B\mu} \text{ . rad mt r}^2\text{/weber}$$

Where 'φ' is the hall angle and

$$\varphi = \frac{V_H}{V_x} \frac{Lx}{b}$$

Lx and b are in metres

Lx is the length and 'b' is breadth of semiconductor crystal.

Basic Principle Involved. When a current carrying conductor placed in a transverse magnetic field, the magentic field exerts a deflecting force F on the charge carriers and given by $F = e(V_D \times B)$, in the direction perpendicular to both drift velocity V_D and applied from one surface to other and as a result excess positive charge at one surface develops an electric field. The voltage associated with this electric field is called Hall voltage and effect is known as Hall Effect.

EXPERIMENTAL PROCEDURE

Connect the circuit and follow the process as given below :

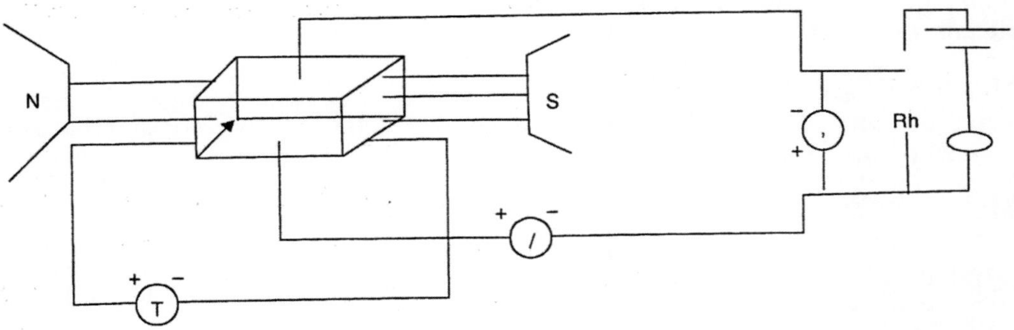

Fig. 13.1.

1. Apply magnetic field along thickness of the crystal and measure the magnetic field with a gauss meter or flux meter and find actual field in the crystal

   $B_\mu = \mu B$ where 'μ' is the permeability of the crystal.
2. Pass the current I_x through the semiconductor crystal along length and measure. Hall Voltage V_H with the help of milli voltmeter along breadth of the crystal. Aslo V_x is measured by voltmeter.
3. Change the current I_x by rheostat R*h* and note down corresponding values of V_x and V_H and take the following observations.

OBSERVATIONS

Width of the crystal d = ... metre
Breadth of the crystal b = ... metre
Length of the crystal L = ... metre
Magnetic field B_μ = ... μB gauss
Least count of voltmeter = ...
Least count of millivoltmeter = ...
Least count of ammeter = ...

S.No.	Current I_x (ampere)	Voltage V_x (volt)	Hall voltage V_H (millivolt)	$\frac{V_H}{I_x}$ (milli ohm)	$\frac{V_H}{V_x}$
1.					
2.					
3.					
4.					

Mean $\frac{V_H}{I_x}$ = Mean = $\frac{V_H}{V_x}$.

CALCULATION

Hall coefficient $R_H = \frac{V_H}{I_x} \times \frac{d}{B\mu} \times 10^4$ ohm-metre/Weber

$$N = -\frac{1}{R_H e} = ...$$

$\mu = \varphi$. Radian-metre^2/weber

$B\mu = \mu B.$

Maximum probable error

$$R_H = V_H d \times 10^4$$

Taking log of both the sides and differentiating

$$\frac{dR_H}{R_H} = \frac{dV_H}{R_H} + \frac{d(d)}{d} - \frac{d(Ix)}{Ix} - \frac{d(B_\mu)}{B_\mu}$$

For maximum error :

$$\frac{dR_H}{R_H} = \frac{d\,V_H}{R_H} + \frac{d(d)}{d} - \frac{d(Ix)}{Ix} - \frac{d(B_\mu)}{B_\mu}$$

$$dR_H = \left[\frac{dV_H}{V_H} + \frac{d(d)}{d} - \frac{d(Ix)}{Ix} - \frac{d(B_\mu)}{B_\mu}\right] R_H$$

Where dV_H = L.C. of millivoltmeter

$d(d)$ = L.C. of scale used to measure 'd'

$d(Ix)$ = L.C. of ammeter

$d(B_\mu)$ = L.C. of gauss meter or flux meter

Similarly

$$\frac{dn}{n} = \frac{dR_H}{R_H}$$

$$\frac{d\mu}{\mu} = \frac{dV_H}{V_H} + \frac{d(Lx)}{Lx} + \frac{d(b)}{b} + \frac{d(Bz)}{Bz}.$$

RESULT

Hall coefficient $R_H = \ldots\ldots \pm dR_H$ ohm-m^3/weber

Carrier's concentration $n = \ldots\ldots \pm dn$ metre^3

Mobility $\mu = \ldots\ldots \pm d\mu$ rad . metre^2/weber.

SOURCES OF ERROR AND PRECAUTIONS

1. Hall voltage should be measured along the direction perpendicular to both the direction of B and Ix.
2. Care should be taken to measure the reading of milli voltmeter.
3. Ammeter reading should be increased in constant proportion.

VIVA-VOCE

Q.1. What is Hall Effect ?

Ans. When a current carrying conductor is placed in a direction perpendicular to the direction of current, a potential difference is developed at the right angles to both current and magnetic field. This phenomenon is called Hall Effect.

Q.2. What is the importance of Hall Effect ?

Ans. By Hall Effect we can calculate

(*a*) Magnetic field

(*b*) We can decide whether a specimen is a semi conductor, conductor or insulator.

(*c*) Mobility of charge can be measured.

(*d*) Nature of charge carriers can be determined from the sign of the hall potential difference.

Q.3. Why does Hall-potential difference arise when a transverse magnetic field is applied to a current carrying conductor ?

Ans. When a current carrying conductor is placed in a transverse magnetic field, the magnetic field exerts a deflecting force $F = e(V_d \times B)$ in a direction perpendicular to both V_d and B. Hence charge are drifted by this deflecting force and as a result excess positive and negative charges on opposite surfaces of one another create an electric field and thus a potential difference is developed.

Q.4. How do you define Hall coefficient ?

Ans. Hall coefficient R_H is numerically equal to Hall electric field induced in the specimen crystal by a unit current when it is placed perpendicular to a magnetic field of 1 weber/m^2.

Q.5. What is the unit of the Hall coefficient ?

Ans. ohm – m^3/weber.

Q.6. What is mobility of charge carriers ?

Ans. It is the ratio of average drift velocity of the charge carriers to the applied electric field.

Q.7. How does mobility depend upon electrical conductivity of the specimen ?

Ans. Mobility of the charge carriers increases with increase of the electrical conductivity of the specimen.

Q.8. What is the unit of mobility ?

Ans. m^2/volt-sec or rad-mtr^2/weber.

Q.9. What is the hall angle ?

Ans. Electron or hall *e*-is under the influence of applied electric field E*x* and instantaneously applied hall electric field E_H at right angles to one another the angle made with *x*-direction by drift velocity U_x of charge carriers is known as Hall angle.

$$\phi = \frac{E_H}{E_x}.$$

Q.10. What type of the charge carrier has greater mobility ?

Ans. In semiconductor, electrons have great mobility.

Q.11. What happens to the hall coefficient, when number of charge carriers per unit volume (*n*) is decreased ?

Ans. R_H is inversely proportional to '*n*', hence R_H is decreased.

14

Energy Band Gap by Four-Probe Method

EXPERIMENT

OBJECT

To study variation of resistivity of the Ge sample (semiconductor) with temperature and to determine the energy band gap by 4 probe method.

APPARATUS REQUIRED

Oven, thermometer, Ge-sample (semiconductor) and 4 prove set up.

FORMULA USED

Since $\rho_0 = \dfrac{V}{I} \times 2\pi s$ where ρ_o = resistivity of crystal

$= \dfrac{V}{I} \times 2 \times 3.14 \times s$ S = distance between the probes.

Since thickness of the crystal is small compared to the probe distance a correction factor for it has to be applied. Further, the bottom source is non-conducting in the present case.

Then, $\rho_G = \dfrac{\rho_0}{G_7\left(\dfrac{w}{s}\right)}$ where w = thickness of the crystal

the function $G_7(w/s)$ may be obtained from table 1.

Hence $\rho_G = V/I \times 2 \times \pi \times s/G7(w/s)$

Thus ρ_G may be calculated for various temperatures.

Plot a graph for

$$\text{Log}_{10}\, \rho_G \ vs\ \frac{1000}{T(K)}$$

Slope of the curve is given by

$$\frac{\log_e \rho}{\frac{1}{T}} = \frac{E_g}{2kT}$$

where k = 8.6×10^{-5} ev/deg and is called boltzmann constant.

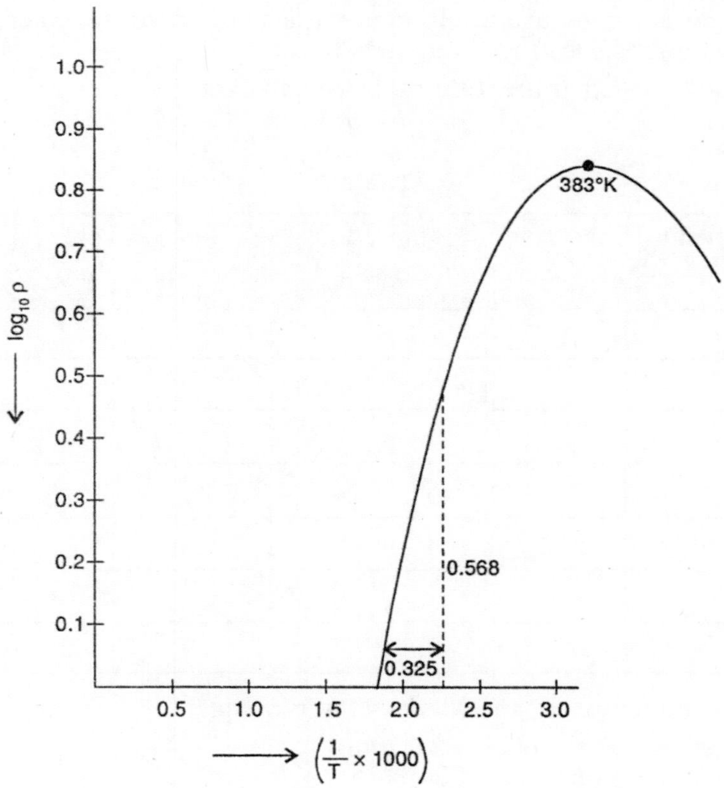

Fig. 14.1.

Table 1

w/s	0.100	0.141	0.200	0.333	0.500	1.000	1.414	3.333	5.000	10.000
G6(w/s)	0.0000019	0.00018	0.00342	0.6014	0.228	0.683	0.848	0.9838	0.9948	0.9993
G7(w/s)	13.863	9.704	6.931	4.159	2.780	1.504	1.223	1.0228	1.0070	1.00045

EXPERIMENTAL PROCEDURE

1. Press the Ge crystal in the 4 probe.
2. Connect the set up with the mains and switch it on.

3. Put the selector switch in voltage position and range multiplier switch as X_I position. Adjust the 0 in DVM with the help of ZERO ADJ keeping either the current terminal in open or voltage terminal in short.
4. Change the selector switch in current position and apply some current say 2mA constant for whole set of readings and note this value of current. Change the selector switch in voltage position at X 10 position and note the corresponding voltage in DVM and temperature.
5. Record these readings in table.
6. Put the oven switch at L position. Connect the oven with the oven supply and switch ON. A glowing LED will show the ON position of the supply. Measure the voltage at different values of temperature keeping the current constant. Record these readings in table.
7. Repeat the experiment for other values of current.

OBSERVATIONS

Table 2

S.No.	*Temperature* $T\,^\circ K$	*Voltage* *V volt*	*Temperature* $T\,^\circ K$	ζ *ohm-cm* *cm*	$1/T \times 10^{-3}$ T	*log* 10ζ
1.						
2.						
3.						
4.						

Current I = *m*A (const for the whole set of readings)

Distance between the probes = 2.0 mm

Thickness of crystal (w) = 0.5 mm.

CALCULATION

Calculation will be done by graph of $\log_{10} \rho_G$ *vs* 1000/T(K).

RESULT

1. Table 1 shows the variation of the Ge sample with temperature.
2. Graph shows the variation of resistivity of Ge sample with temperature.
3. The experimentally observed value of Eg = eV.
4. The standard value of Eg = 0.70eV.

SOURCES OF ERROR AND PRECAUTIONS

1. Their should be 0 reading at DVM initially.
2. All the 4 probes should be well in touch with the Ge crystal.
3. The current should be taken constant throughout the coarse of the experiment for a particular observation.
4. Care should be taken to compute resistivity.

VIVA-VOCE

Q.1. What is semi-conductor diode ?

Ans. A *p*-type semi conductor in contact with an *n*-type semiconductor constitutes a *p-n* junction diode.

Q.2. What do you mean by energy band gap ?

Ans. In fact the energy level of semiconductors has a band structure as shown in the figure below :

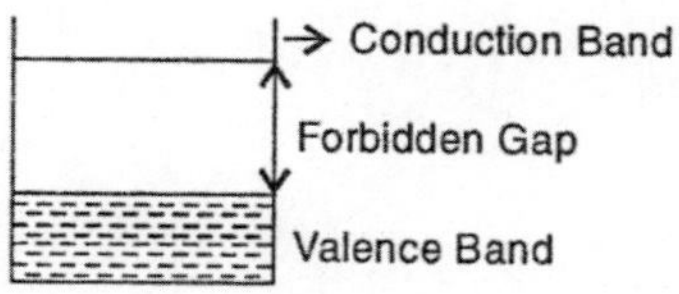

Fig. 14.2.

The band which is occupied by valence electrons is called valence band. It may be partially or completely filled.

The band which is almost empty to electrons is called conduction band.

The gap between conduction band and valence band is energy band gap and called forbidden gap.

Q.3. How are Conductors, Insulators and Semiconductors differentiated on the basis of energy band gap ?

Ans. In a good conductor conduction band and valence band overlap and hence no band gap will be there.

In insulator the energy band gap is about 7eV while in semiconductor it lies between 0.7eV and 1.1 eV.

Q.4. What are intrinsic and extrinsic semiconductors ?

Ans. In intrinsic semiconductors, number of free electrons and holes are solely created by thermal extiation and number of holes and electrons are equal. Pure Ge and Si are example of intrinsic semiconductor. It has small electrical conductivity. When a pentavalent impurity (Sb) is added to intrinsic semiconductor (Ge or SiO), number of free electrons increases and when a trivalent impurity (Al) is added, number of holes increases. Thus, conductivity would be increased and semiconductor is called extrinsic semiconductor. The former extrinsic semiconductor posses excessive's and called *n*-type while later posses positive holes and called *p*-type.

Q.5. What is depletion layer ?

Ans. When *p-n* junction diode is formed, hole from *p*-region diffuses into *n*-region and electrons from *n*-region diffuse into *p*-region. During the formation of the junction *e*, *s*-holes recombination takes place and as a result a thin layer at junction, free from charge carriers is formed. This region is called depletion region. Its thickness is of order 10^{-6} m.

Q.6. What is meant by potential barrier in p-n junction ?

Ans. During formation of *p-n* junction diode a deplection layer is formed and it is separated by 2 layers, one of immobile + ve ions on *n*-side and immobile –ve ions on *p*-side of the junction and called potential barrier.

Q.7. What type of oil do you use ? Why not water ?

Ans. Oil has high boiling pt. whereas water has boiling pt. of only 100°C.

Q.8. Why 4 probe method is better than other methods for measuring the resistivity ?

Ans. In this method the current carrying contacts inject minority carriers which changes the resistance of the metal.

15

Flat Spiral Spring

EXPERIMENT

OBJECT

To determine the force constant of a flat spiral spring by statical method and to verify series and parallel combination theorem.

APPARATUS

Spiral spring, weight, stopwatch.

FORMULA USED

For Statical method

$$K = \frac{mg}{x}$$

Where

m = mass suspended from the spring

x = corresponding extension of the spring.

EXPERIMENTAL PROCEDURE

1. Suspend the spring from a rigid support. Put a scale just behind the pointer which is attached to the lower end of the spring. Note down the position of the pointer.
2. Hang a suitable weight from the lower end of the spring and note down the position of the pointer. Go on increasing the weight hanged in suitable steps and corresponding to each weight read position of the pointer.
3. Now decrease the weight in the same steps and again note down the position of pointers in each step.
4. Calculate the mean extension corresponding to each weight and then K as given in the table.
5. Draw a graph between M(on X-axis) and extension x(on Y-axis). If θ is the inclination of the straight line then K is given by $K = g \cot \theta$.

6. Repeat above steps for series as well as parallel combination of the spring and determine equivalent force constant.

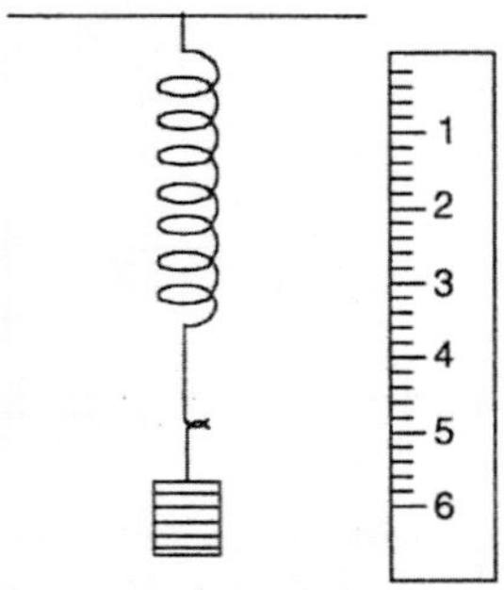

Fig. 15.1.

OBSERVATIONS

Mass of the pan M_0 = ... gm

S.No.	*Load M(gm)*	*Reading of the Load Increasing cm (a)*	*Pointer Load Decreasing cm (b)*	*Mean Extension (a + b)/2*
1	0			
2	200			
3	300			
4	400			
5	.			
.	.			
.	.			
.	.			

CALCULATION

Plot graph between mass M and the extension x. A linear plot as shown in the Fig. 15.2 is obtained. The slope of this straight line will yield the spring constant K.

$$K = \frac{(M_2 - M_1)g}{x_2 - x_1} \text{ Nm}^{-1}$$

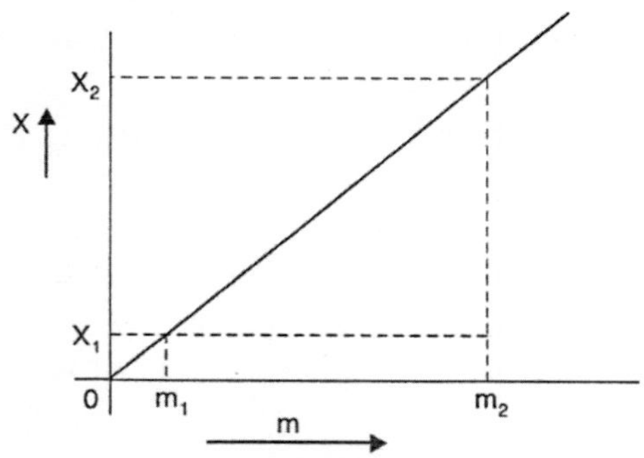

Fig. 15.2.

Maximum probable error

$$\delta k = \frac{2\delta M}{(m_2 - m_1)} + \frac{2\delta x}{x_2 - x_1}$$

Where,

δm = mass corresponding to one small division of the graph

δx = extension corresponding to one small division of the graph.

RESULT

The value of the force constant of the given spring = ± .... Nm^{-1}

SOURCES OF ERROR AND PRECAUTIONS

1. Mass should be put gently on the pan. Wait for few minutes before recording the reading for extension.
2. While loading the spring, the extension should always be with in the elastic limit.
3. Reading should be taken of both increasing and decreasing loads.

16

Diffraction Gratting

EXPERIMENT

OBJECT

To determine the wavelength of light emitted by a given source (Hg Vapour Lamp) with diffraction gratting.

APPARATUS REQUIRED

Spectrometer, diffraction gratting, Mercury vapour lamp, prism and reading lens.

FORMULA USED

$$\lambda = \frac{g_e \sin\theta}{n} = \frac{(a+b)\sin\theta}{n}$$

Where,

λ = wavelength of the spectral line

g_e = gratting element

n = order of spectrum

θ = angle of diffraction.

EXPERIMENTAL PROCEDURE

1. *Following preliminary adjustment should be made before using the spectrometer*
   (*a*) The axis of the telescope and that of the collimator must intersect the axis of instrument.
   (*b*) Prism table should be levelled.
   (*c*) The telescope and the collimator must be set for parallel light beam by Schuster's Method.
2. *Setting of Gratting normal to the incident light.*
   (*a*) Arrange the telescope and the collimator in a line so that the direct image of the slit coincides with the vertical cross wire. Let the vernier reading be ϕ.

(*b*) Mount the gratting on the prism table with the help of clamping device such that its plane at the centre edge is perpendicular to the two levelling screw A and B.

(*c*) Rotate the telescope through 90° *i.e.*, to the position (90 + ϕ). Now rotate the prism table till the image of the slit falls on the intersection of the cross wire. If the image is not in the centre, it can be shifted to the centre with the help of screw A and B. In this position, the plane of the gratting is inclined at 45° to the incident light as shown in the Fig. 16.1. Note down the vernier reading.

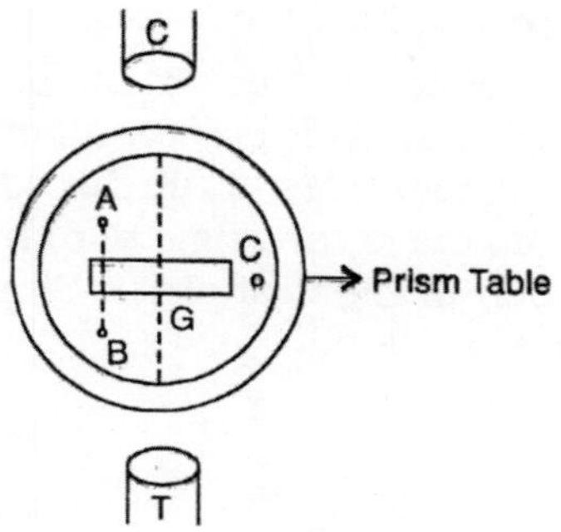

Fig. 16.1.

(*d*) Now turn the prism table from this position through 45° or 135°. In this position of the gratting is normal to the incident light. Clamp the prism table.

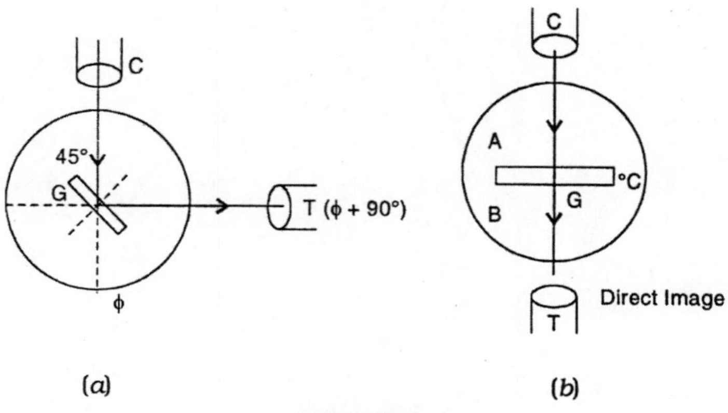

Fig. 16.2.

3. *Setting of the slit parallel to the lines of the gratting*

Rotate the telescope either to left or to right to get a first order spectral line on the cross-wire. Rotate the slit in its own plane so that spectral line may become sharp. If necessary, the third screw C may be used to sharpen the line (image).

4. *Measurement of anlge of Diffraction*

(*a*) The telescope is rotated about the axis of the instrument to the left side of the direct image until the first order spectrum is in view. With the help of tangent screw, the telescope is moved till its vertical cross-wire coincides with one of the spectral line (say violet). The telescope is now clamped and the readings of both the vernier is noted. Unclamp the telescope, move it so that the vertical cross-wire falls on the next spectral line (say yellow), and note down the reading of both the verniers. Similar readings are recorded for other spectral lines.

(*b*) Move the telescope to the first line of the second order. Coinciding the cross-wire on different spectral lines, the corresponding vernier readings are noted.

(*c*) Now move the telescope to the right of the direct image and similar observations are taken in the first and the second order. The difference of the same vernier reading on the two sides of the direct image of the same spectral line in the first order gives twice the angle of diffraction. Half of this angle θ of is the angle of diffraction in the first order. Similarly the angle of diffraction in the second order is calculated.

OBSERVATIONS

Least count of Vernier = ... sec

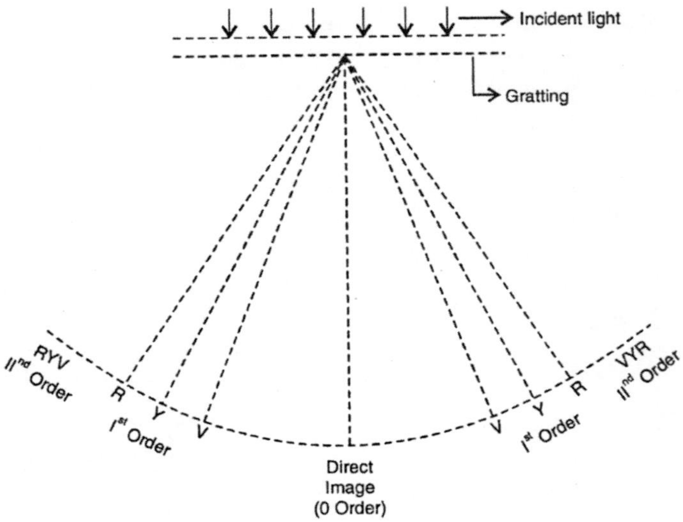

Fig. 16.3.

S.No.	Order	Colour	Verniers	Left of direct image			Right of direct image				
				M.S.	V.S.	Total (a)	M.S.	V.S.	Total (b)	2θ	θ
(1)		V	V_1 V_2								
	I	G	V_1 V_2								
		Y	V_1 V_2								
(2)	II	V	V_1 V_2								
		G	V_1 V_2								
		Y	V_1 V_2								

CALCULATION

Total No. of Lines per inch N_0 = ... per inch

Grating Element $g_e = 2.54/N_0$ per cm

First Order ($n = 1$)

Violet $\lambda v = g_e \sin \theta v$

= .... m

Green $\lambda g = g_e \sin \theta g$

= .... m

Yellow $\lambda y = g_e \sin \theta y$

= .... m

Second Order = ($n = 2$) .... m

Violet $\lambda v = \dfrac{g_e \sin \theta v}{2}$ = ... Å

Similar calculation may be made for other colours.

Maximum Probable Error

$$\delta\lambda = (\cot \theta \, \delta \theta) \, \lambda$$

RESULT

The wavelengths of the yellow, green and violet colours are shown in the table.

Colours	Measured (mm)	Standard (mm)
Yellow-I	...............±	579.06
Yellow-II	...............±	576.96
Green	...............±	546.07
Violet	...............±	435.83

SOURCES OF ERROR AND PRECAUTIONS

1. Preliminary adjustments of the spectrometer are most essential.
2. Gratting should be set normal to the incident light.
3. Gratting should not be touched in the middle with fingers.
4. While taking observations, the prism table and then telescope should be clamped and the readings of both the verniers should be noted.

17

Lee and Charlton Method for Bad Conductors

EXPERIMENT

OBJECT

To determine the thermal conductivity of bad conductor (asbestos) by Lee and Charlton method.

APPARATUS REQUIRED

Lee and Charlton apparatus, heating arrangement, boiler, stopwatch, vernier callipers, screwgauge.

FORMULA USED

$$K = \frac{msb}{A(\theta_1 - \theta_2)} \times \left(\frac{\Delta\theta}{\Delta T}\right)_{\theta_2}$$

Where,

m = mass of the lower disc

s = specific heat of the material

$A = \pi R^2$ = area of the top surface of the lower disc.

B = thickness of the specimen

θ_1, θ_2 = the steady state temperature of two faces of the specimen

$\left(\frac{\Delta\theta}{\Delta t}\right)_{\theta_2}$ = slope of the cooling curve at temperature θ_2.

EXPERIMENTAL PROCEDURE

1. Measure the thickness d of the specimen. Determine the mass m and the radius of the lower disc. Sandwich the specimen between the team chamber and the lower disc. Insert the thermometers T_1 and T_2 into the holes provided for them.
2. Suspend the assembly from threads and allow the steam to pass through the steam chamber. Wait till steady state is reached. In the steady state, the thermometers do not register any change in temperature θ_1 and θ_2.

3. Take out the lower disc and suspend it alone. Heat it till its temperature rise about 10°C above θ_2 and then put the specimen over its top surface. Allow to cool the disc till its temperature falls 10°C above θ_2 and during the cooling process, note down the temperature every minute or half-minute. Draw cooling curve by plotting time t on X-axis and temperature θ on Y-axis. Draw a tangent at the point corresponding to the temperature θ_2. The slope of the tangent represents.

OBSERVATIONS

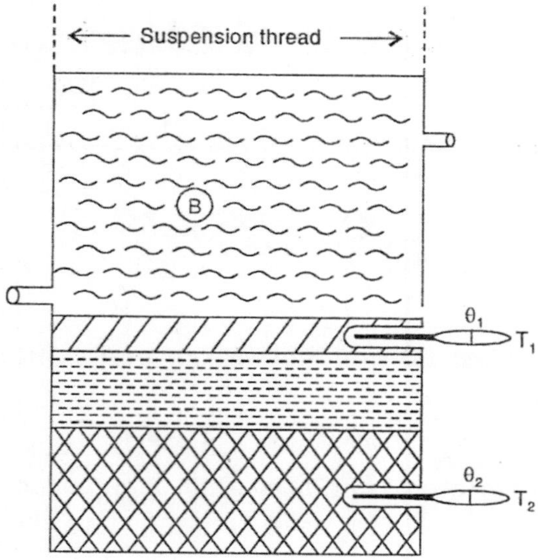

Fig. 17.1. Lee's Charlton Apparatus.

Mass of the disc m = ... gm

Diameter of the disc d = ... cm

Area (A) = $\pi d^2/4$ = ... cm^2

Thickness of the specimen = ... cm

Specimen Heat S = ... K cal kg^{-1} s^{-1}.

Observations for cooling of the disc

S.No.	Time (in sec)	Temperature of Disc
1.	0.0	
2.	20	
3.	60	
4.	90	
5.	.	
6.	.	
.	.	
.	.	
.	.	

CALCULATION

Convert all the measured quantities into S.I. unit. Find out the value of

$$\frac{\Delta\theta}{\theta T} = \frac{\theta_3 - \theta_2}{t_2 - t_1}$$

Put the value of different quantities in the formula and evaluate the value of K.

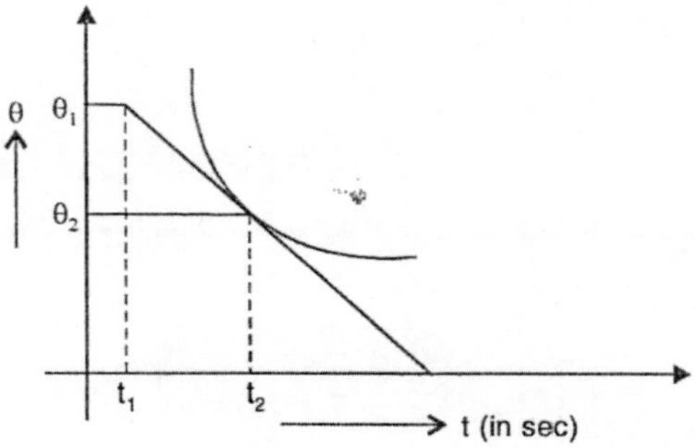

Fig. 17.2.

Maximum Probable Error

$$K = \frac{4msb}{\pi d^2(\theta_1 - \theta_2)} \times \left(\frac{\Delta\theta}{\Delta t}\right)_{\theta_2}$$

$$\delta K = \left[\frac{\delta m}{m} + \frac{\delta b}{b} + \frac{2\delta d}{d} + \frac{\delta\theta_1 + \delta\theta_2}{\theta_1 - \theta_2} + \frac{\delta\Delta\theta}{\Delta\theta} + \frac{\delta\Delta t}{\Delta t}\right] K$$

Where $\delta(\Delta\delta)$ is the value of one small division in the graph on θ-axis, $\delta(\Delta t)$ is the value of one small division in the graph on t-axis and all the δ's area least count of different instrument used.

RESULT

The thermal conductivity of the medium is ± .... $Js^{-1}m^{-1}(^oC)^{-1}$.

SOURCES OF ERROR AND PRECAUTIONS

1. In order to make good contact of the specimen with the steam chamber and the lower disc, a little amount of Glycerine should be smeared over the metallic surfaces.
2. The cross-section of the specimen should be equal to that of the lower upper discs.
3. θ_1 and θ_2 should be recorded in the steady state.
4. The thickness of the specimen should be measured at different points and mean should be taken.
5. The formula for thin specimen because in the derivation the heat lost from its current surfaces has been neglected.

18

Helmholtz Galvanometer

EXPERIMENT

OBJECT

To determine the reduction factor of Helmholtz galvanometer using copper Voltameter.

APPARATUS REQUIRED

Helmholtz galvanometer, copper voltameter, rheostat, four way key, chemical balance and a cell.

FORMULA USED

The current flowing in the Helmholtz galvanometer is given by

$$i = K \tan \theta, \qquad K = \frac{10(\sqrt{125})aH}{32n\pi}$$

And a current flowing in a copper voltameter is given by

$$i = \frac{m}{zt}$$

$$\therefore \qquad \frac{m}{zt} = K \tan \theta$$

$$\text{or} \qquad K = \frac{m}{zt \tan \theta}$$

K = reduction factor of Helmholtz galvanometer,

a = radius of the coil,

H = horizontal component of the earth's magnetic field in gauss,

θ = deflection in the Helmholtz galvanometer,

n = number of turns of wire in the coil,

m = mass of the copper deposited on cathode of the copper voltameter,

z = electro-chemical equivalent of copper, and

t = time for which current is passed in the copper voltameter.

EXPERIMENTAL PROCEDURE

The connection is made to be view of Fig. (18.1). Take care in connecting the Helmholtz galvanometer. Each coil has three or four points at its base representing the number of turns. Connect the second or third points of the two coils with the help of a connecting wire. Note the corresponding number of the turns of the wire. The first point of the two coils will serve the purpose of '*a*' and '*b*' of the Fig. (18.1). Set the plane of the two coils parallel to the small magnet of the compass box placed between the two coils. On achieving this condition the coils are close to the magnetic meridian. Now connect the points 1 and 2 respectively to 3 and 4 of the four ways and put the test plate as cathode in the copper voltameter. Switch on the circuit and adjust the rheostat R so that the galvanometer reads deflection of 45° now reverse the direction of current by connecting the points 1 and 3 respectively to 2 and 4 of the fourth way key. If the deflection in the galvanometer is again 45° then the coils are in the magnetic meridian otherwise rotate the planes of the coils in such a way that the same deflection is observed on reversing the current as in the direct one.

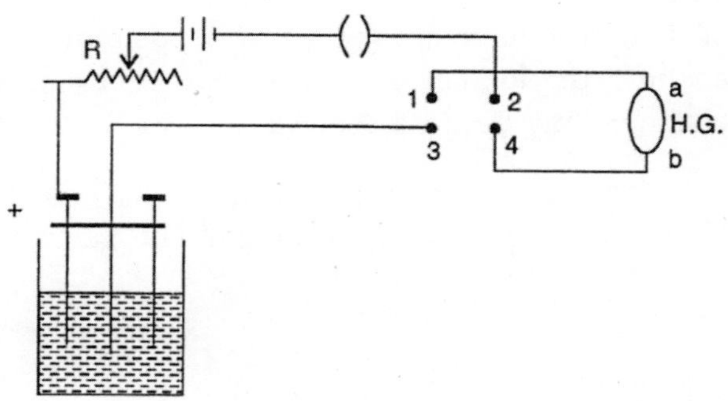

Fig. 18.1.

Switch off the circuit by taking out the plug of one-way key and replace the test plate by clean and weighed copper plate. Connect the points 1 and 2 and 3 and 4 of the four way key. Allow the current to flow in the circuits by inserting the key and simultaneously start the stopwatch. After 15 or 20 min, the current in reversed with the help of four way key and is passed for the same time as in the direct one. It will be beneficial to record the deflection after every 5 min. Note the total time for which the current is passed. The negative plate of the copper voltameter, on which cooper has been deposited, is taken out and washed carefully in a beaker containing clean water. It is then dried and weighed on a chemical balance.

OBSERVATIONS

The electro-chemical equivalent of the copper

$Z = 0.0003295$ gm/coulomb

Number of the turns of the coils = ...

Radius of the coils = ...

1. Deflection in galvanometer

S.No.	Time minutes	Deflection				Total time t sec	Mean deflection θ
		Direct current		Reverse current			
		One end of pointer °	Other end of pointer °	One end of pointer °	Other end of pointer °		
1.	5						
2.	10						
3.	15						
4.	20					1800	
5.	25						
6.	30						

2. Mass deposited :

(*i*) Mass of the cathode before copper is deposited = ... gm
(*ii*) Mass of the cathode after copper is deposited = ... gm
(*iii*) Mass of the copper deposited m = ... gm.

CALCULATION

Mean value of the radius of the coil of the galvanometer a = ... cm

$$K = \frac{m}{zt \tan \theta} = \text{.......} \text{ amp}$$

Theoretical value of the reduction factor

$$K = \frac{10\sqrt{(125aH)}}{32n\pi} = \text{.......} \text{ amp.}$$

Probable Error

$$K = \frac{m}{zt \tan \theta}$$

Take logarithm of the both sides of this equation and differentiation to obtain, for the maximum probable error

$$\frac{dK}{K} = \frac{dm}{m} + \frac{dz}{z} + \frac{dt}{t} + \frac{\sec^2 \theta}{\tan \theta} d\theta$$

dm = sensitivity of the chemical balance

dz = 0

$d\theta$ = L.C. of the magnetic compass expressed in radian

= L.C. of the magnetic compass in degree $x\,\pi/180$

Using the values of the various quantities here calculate dK.

RESULT

The reduction factor for .... number of turns of the coils of the Helmholtz galvanometer = ... ampere.

SOURCES OF ERROR AND PRECAUTIONS

1. Reading of the magnetic compass should be taken by placing the eye above the position of the pointer collinear with its image in the mirror provided at the base of the compass.
2. There should not be any magnetic material in the vicinity of the apparatus.
3. Check that negative of the circuit is connected to the cathode of the copper voltameter.
4. Do not dry the copper plate in the direct sunlight.
5. Coils should be in the magnetic meridian otherwise tangent law would not hold good.

VIVA-VOCE

Q.1. What do you mean by reduction factor of Helmholtz galvanometer ?

Ans. The reduction factor of Helmholtz galvanometer is equal to the amount of current in amperes which produces a deflection of 450 in the galvanometer placed in the magnetic meridian.

Q.2. What is its unit ?

Ans. Its unit is amperes.

Q.3. What are the factors on which the reduction factor of Helmholtz galvanometer depends ?

Ans. The reduction factor of Helmholtz galvanometer depends on

(*i*) The radius 'a' of its coils,

(*ii*) The horizontal component H of the earth magnetic field.

(*iii*) The number of turns 'n' in the coil.

Q.4. What is a tangent galvanometer ?

Ans. The tangent galvanometer is an apparatus which measures current in terms of the deflection in the magnetic compass placed at the centre of the coil through which the current is allowed to flow.

Q.5. Why do we keep the tangent galvanometer in the magnetic meridian ?

Ans. When we place coil of the galvanometer in the magnetic meridian the magnetic field produced at the centre of the coil due to current flowing through in it is normal to the horizontal component of earth's magnetic field.

Q.6. How does Helmholtz galvanometer differs from tangent galvanometer ?

Ans. Helmholtz galvanometer is modified form of tangent galvanometer. In a tangent galvanometer the magnetic field produced by this does not remain uniform around the centre whereas in Heltmoltz galvanometer the magnetic field produced remains uniform around the centre of the galvanometer. To achieve this there are two coils separated by a distance equal to the radius of the coils and the magnetic compass is placed in between the coils equidistant from them.

19

Potentiometer

EXPERIMENT

OBJECT

(*a*) ***To calibrate an ammeter and a voltmeter, and***

(*b*) ***To measure a given low resistance by potentiometer.***

APPARATUS

Potentiometer, standard cell, rheostat, one way key, two way key, two cells, one ohm resistances and the decimal ohm box.

FORMULA USED

The small resistance (r) is given by

$$r = \frac{R(l_1 - l_2)}{l_2}$$

Where, R = resistance of decimal ohm box, l_1 and l_2 are the positions of null points.

EXPERIMENTAL PROCEDURE

1. **Calibration of voltmeter.** An ideal voltmeter, in principle, has an infinite resistance but it cannot be achieved in practice. The voltmeter commercially available is constructed by putting a high resistance in series with a galvanometer. Therefore, the reading of voltmeter connected in an electrical circuit will not give the true value of the potential difference across the points connected. In order to know the amount of error one has to calibrate it. And this can be easily done with the help of potentiometer as it gives the true value of the potential difference. For this purpose we use of an auxiliary circuit as shown in Fig. 19.1, and insert it at the appropriate place. We measure the potential difference across the two points X and Y (*i*) by voltmeter and (*ii*) by potentiometer. Let V be the reading of the voltmeter and ΔV be the difference of the reading from one of the result obtained by potentiometer. ΔV represents the error of the voltmeter reading V and it may be positive as well as negative. A graph between V and ΔV known as error graph can be

plotted. There is no exact nature of this graph. This graph may be used to know the amount of error in the reading of the voltmeter.

2. **Calibration of an ammeter.** An ideal ammeter should have in principle zero resistance, which I not possible to achieve. However in constructing an ammeter one uses a low resistance shunt wire in parallel to a galvanometer. In this way, the total resistance of the galvanometer reduces to a lower but in any case infinite. Therefore, reading of an ammeter does not represent the true value of the current. It has therefore, to be calibrated in order to do this we make use of the auxiliary circuit diagram, as shown in the Fig. 19.1. Together with the circuit of Fig. 19.1, we measure the current flowing in the circuit in two ways (*i*) by an ammeter and (*ii*) by determining the potential using the potentiometer across the terminals X and Y of a 1-ohm resistance.

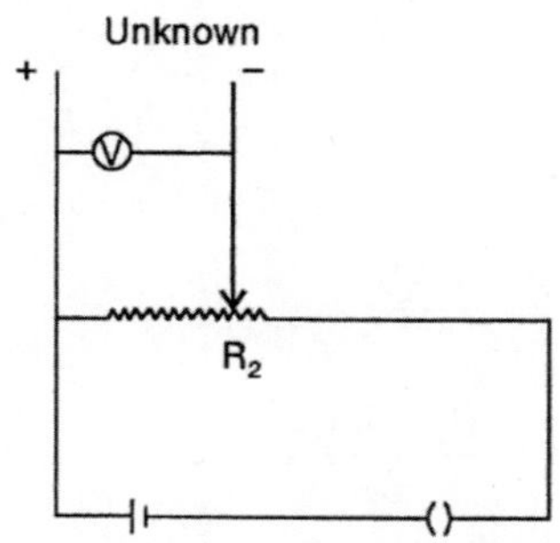

Fig. 19.1. Circuit diagram for calibration of voltmeter.

The potential across the terminals of one-ohm resistance is equal to the current flowing in it. Let I be the reading of the ammeter and ΔI the difference of this reading from one obtain by potentiometer. ΔI represent the error of the ammeter reading I and may be positive as well as negative. An error graph between I and ΔI may be plotted. There is no exact nature of this graph. The graph estimates the amount of error in the reading of an ammeter.

3. **Measurement of low resistance.** The accurate measurement of low resistanc is not possible with the help of conventional apparatuses available in the laboratory. For this purpose, potentiometer is most suitable. We make use of an auxiliary circuit as shown in Fig. 19.2. We choose a standard resistance box such as decimal ohm box represented by R in the circuit and determine the potential differences across (R + *r*) and R, where *r* is the unknown low resistance.

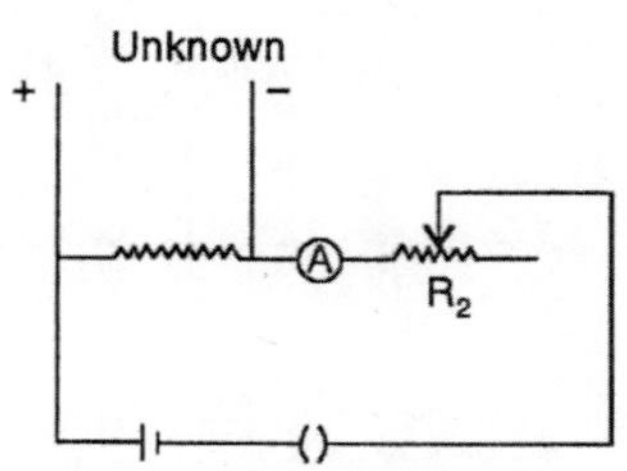

Fig. 19.2. Circuit diagram for calibration of ammeter.

The potential difference V_1 across the resistance $(R + r)$ is given by :

$$V_1 = I(R + r)$$

Where I represent the current flowing in the circuit. The potential difference V_2 across the resistance R is given by

$$V_2 = IR$$

Dividing these equations we get :

$$\frac{V_1}{V_2} = 1 + \frac{r}{R}.$$

We measure the potential difference V_1 and V_2 by potentiometer. Let l_1 and l_2 be the corresponding length for null point, then in view of equation, we have :

$$\frac{V_1}{V_2} = \frac{l_1}{l_2}$$

Hence equation becomes

$$\frac{l_1}{l_2} = 1 + \frac{r}{R}$$

or

$$r = \frac{(l_1 - l_2)R}{l_2}$$

EXPERIMENTAL PROCEDURE

Before making any observation with potentiometer it has to be standardized :

1. **Standardization of potentiometer.** Make the connections of the main circuit as shown in Fig 19.3. Connect the terminals 1 and 3 of the two way key and set the jockey at a point B such that the effective length of the wire becomes 1018.6 cms, in case a cadmium cell is used as standard. Let us assume the length of the small wire CD to be 50.0 cms. Then each coil attached at the point D corresponds to a wire of length 50 cms. Hence, if we use 20 coils and place the jockey at B keeping DB = 18.6 cms, an effective length of 1018.6 cms will be obtained. Now press the jockey at B and adjust the current in the main circuit with the help of rheostat R_1 in order to get the null deflection in the galvanometer. In this way the potentiometer is standardized to have a potential gradiant $K = 10^{-3}$ volt/cm and is ready for any measurement.

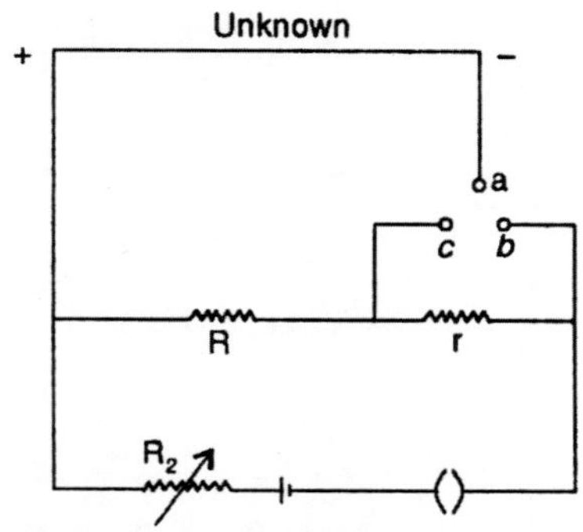

Fig. 19.3. Circuit diagram for determination of low resistance.

2. **Calibration of voltmeter.** First of all complete the connections of the auxiliary circuit as shown in the figure. And connect the terminals of the voltmeter to the unknown gap of the main circuit as shown in the figure. Now ensure that all +ve terminals are connected to the end A of the potentiometer. Disconnect the terminals 1 and 3 and connect 1 and 2 of the two way key of main circuit. Use the rheostat R_2 of the auxiliary circuit and obtain some reading in the voltmeter. The value thus obtained corresponds to the potential difference between X and Y. Now determines this potential difference with the help of potentiometer. If l be the length for which a balance point is obtained then the value of the corresponding potential differenc will be KI = 10^{-3} 1 volt, I be measured in cms. Take several observation for potential difference between X and Y by voltmeter and potentiometer by changing its value with the help of rheostat R_2 of the auxiliary circuit. Draw an error graph between ΔV and V where V stands for voltmeter reading.
3. **Calibration of ammeter.** Make the connections of the auxiliary circuit as shown in the fig. And connect terminals X and Y to the unknown gap of the main circuit as shown in the figure. Always take care of sign while making connection. Disconnect point 1 and 3 and connect 1 and 2 of the two way key adjust some value of current in ammeter with the help rheostat R_2 of the auxiliary circuit. According to ohm's law V = IR, the potential difference across 1Ω coil will be numerically equal to the current flowing through it. Measure the potential difference with the help of potentiometer by locating the null point. If I be such length. Then potential difference will be 10^{-3} I and in principle it is the true current flowing in the circuit. Take several observations by changing the current of the auxiliary circuit with the help of rheostat R_2. Draw an error graph between the error ΔI and I, the ammeter reading.
4. **Measurment of low resistance.** Make the connections of the auxiliary circuit of fig. And connect it in the unknown gap of the main circuit. Disconnect the points 1 and 3 and connect points 1 and 2 of the two way key of the main circuit. Balance the potentiometer by first connecting the point '*a*' and '*b*' and then *a* and *c* of the two way key of the auxiliary. Let the corresponding circuit reading be l_1 and l_2. By changing the value of R obtain several observation for l_1 and l_2 using the formula, *r* can be calculated.

OBSERVATIONS

E.m.f. of the standard cell E = ... volt.
Effective length of the wire L = ... cms
Potential gradient of the potentiometer K = ... volt/cm

1. **Calibration of voltmeter :** Least count of voltmeter = ... volt.

S.No.	Voltmeter reading V (volt)	Potentiometer V' (volt)	$\Delta V = V - V'$ (volt)
1.			
2.			
3.			
4.			
5.			
6.			
7.			
8.			
9.			
10.			

2. **Calibration of ammeter :** Least count of ammeter = ... amp.

S.No.	Ammeter reading I amp	Reading of potentiometer, (position of the point) amp	Value of current read by potentiometer I' amp	$\Delta I = I - I'$ amp
1.				
2.				
3.				
4.				
5.				
6.				
7.				
8.				
9.				
10.				

3. **Measurement of low resistance :**

S.No.	R ohm	l_1 cms	l_2 cms	r cms
1.				
2.				
3.				
4.				
5.				

CALCULATION

$$r = \frac{(l_1 - l_2)}{l_2} R$$

Calculate r for each observation separately and obtain mean value of r.

Probable error. $r = \frac{(l_1 - l_2)}{l_2} R$

Using the standard method, the formula for the probable error is obtained as :

$$\frac{dr}{r} = \frac{dl_1 + dl_2}{l_1 - l_2} + \frac{dl_2}{l_2} + \frac{dR}{R}$$

[$dl_1 = dl_2 = d_1$ least count of the scale attached parallel to wire DC, $dR = 0$]

Hence $dr = \frac{(l_1 + l_2)\, d_1}{(l_1 - l_2)\, l_2}$

or $dr =$ .... ohm.

RESULT

The error graph for voltmeter and ammeter is shown. The value of small resistance $r =$ ...... ohm.

SOURCES OF ERROR AND PRECAUTIONS

1. All connecting wires should be clean and short.
2. After standardization the rheostat R_1 should not be changed and disturbed.
3. The positive ends of the battery and the unknown potential should be connected to the same terminal of the potentiometer wire.
4. The e.m.f. E of the cell used in the main circuit should be greater than the unknown potential to be determined.
6. One should never try to measure the e.m.f. of the standard cell by using a voltmeter.

VIVA-VOCE

Q.1. Explain the working of a potentiometer ?

Ans. See the text.

Q.2. How does the measurement of a potential difference made 'with the help of a potentiometer differ from that of a voltmeter ?

Ans. The measurement of potential difference made with the help of Potentiometer is more accurate than that of a voltmeter because the potentiometer measures the potential difference between any two points of an electrical circuit without drawing any current whereas the voltmeter consumes some current due to its finite resistance and accordingly measures a smaller potential difference this the actual one.

Q.3. What do you mean by calibration of voltmeter ?

Ans. An ideal voltmeter has infinite resistance which can not be achieved in practice, the voltmeter available, commercially, is constructed by putting a high resistance in series in a galvanometer. Therefore, the reading of voltmeter does not represent the true value of the potential difference measured. In order to know the amount of error one needs to calibrate a voltmeter.

Q.4. What do you mean by calibration of ammeter ?

Ans. An ideal ammeter has zero resistance. To achieve this one uses a low resistance shunt wire in parallel to a galvanometer. This reduces the total resistance to very low value but in any case infinite. Therefore the reading of an ammeter does not represent the true value of the current measured. In order to know the amount of error one needs to calibrate an ammeter.

Q.5. How do you calibrate an ammeter using potentiometer ?

Ans. We measure the current flowing in a circuit by two ways (*i*) by the ammeter and (*ii*) by determining the potential difference across the terminals of a standard one ohm resistance using potentiometer. The potential difference measured across the terminals of one ohm resistance wire equals the current flowing in the circuit. Thus we calibrate the ammeter.

Q.6. What is the purpose of standardization of a potentiometer, and how do you achieve it ?

Ans. The purpose of standardization is to develop a known potential difference across the unit length of the potentiometer wire. This is achieved by balancing the known *e.m.f.* of a standard cell across the known length of potentiometer wire. In order to get this condition we adjust the current flowing in the potentiometer wire with the help of the rheostat connected in the circuit.

Q.7. What is the accuracy of the measurement of the potential difference measured by a potentiometer ?

Ans. The accuracy is determined by the potential drop across smallest measurable length of the potentiometer wire. Usually we set it to be, 104 volt/mm.

Q.8. What is the largest value of potential difference measurable by the potentiometer ?

Ans. The largest value of potential difference is equal to the potential difference across the total length of the potentiometer wire.

Q.9. Why do you measure small resistance ? Can't one measure I high resistance ?

Ans. Because null print will be found beyond the length of potentiometer wire.

Q.10. Has calibration curve any specific shape ?

Ans. No.

Q.11. In order to measure e.m.f. of a cell which of the two would you prefer (*a*) poentiometer or (*b*) voltmeter and why ?

Ans. Potentiometer, because it takes reading at null deflection *i.e.*, zero potential difference across the terminal of potentiometer.

Q.12. Suppose during standardization one is getting deflection in the galvanometer to only one side. What probable reason (would you ascribe) ?

Ans. Potential-gradient will have to be adjusted.

20

Kater's Pendulum

EXPERIMENT

OBJECT

To determine the value of 'g' (acceleration due to gravity) by Kater's Pendulum.

APPARATUS REQUIRED

Kater's Pendulum, stop watch, measuring scale.

FORMULA USED

$$g = 4\pi^2 L/T^2$$

where L is the distance between two knife edges and T is the time period.

EXPERIMENTAL PROCEDURE

1. The 2 masses M_1 and M_2 as shown in the Fig. 20.1.
2. Put one of the knife edges near to the masses on the pivot point.
3. Allow the pendulum to oscillate. Note down the time T of 50 oscillations.
4. Reverse the pendulum without changing the position of the two masses. Pivot the second knife edge and allow the pendulum to oscillate. Note down the time T_2 of 50 oscillations.
5. If T_2 is not equal to T_1, adjust the distance between the two knife edges, by trial, such that $T_2 = T_1$.
6. Using the value of $T = T_1/50 = T_2/50$ calculate 'g'.
7. Change the position of masses and repeat the above procedure.
8. Take at least 3 readings and calculate g from each reading.
9. Find the average value of g.

OBSERVATIONS

Least count of stopwatch =

S.No.	*Distance between knife edges (cm)*	*Time for 50 oscillations (secs)*	*Time period (secs)*	*Value of 'g' (cm/s²)*
1.				
2.				
3.				

RESULT

Average value of 'g' = cm/s^2.

SOURCES OF ERROR AND PRECAUTIONS

1. The amplitude of oscillations should not exceed 5°.
2. The time period should be noted very carefully.

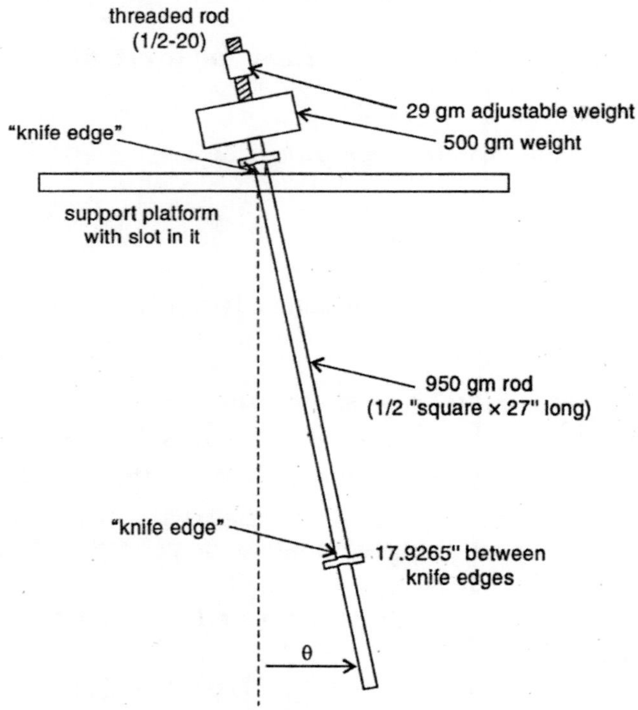

Fig. 20.1.

The period for this arrangement can be proved to be the same as that of a simple pendulum whose length L is the distance between the two knife edges.

KATER'S PENDULUM DETERMINATION OF 'g'.

THEORY

O and P are two knife edges. C is the centre of mass. When swung from O the period is :

$$T_0 = 2\pi\sqrt{\frac{I_0}{mgh_0}} = 2\pi\sqrt{\frac{I_C + mh_0^2}{mgh_0}}$$

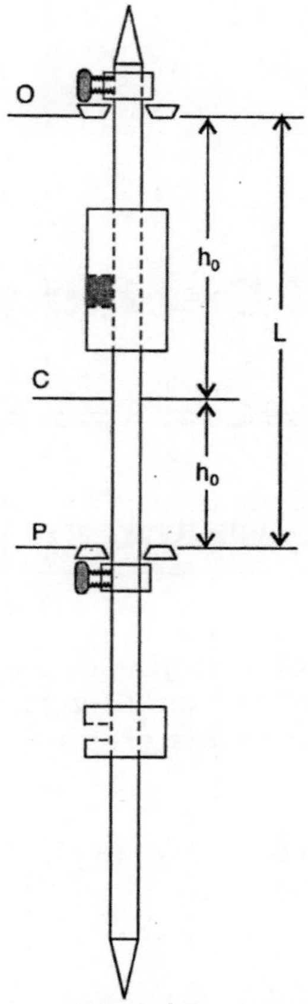

Fig. 20.2.

When swung from P :

$$T_P = 2\pi\sqrt{\frac{I_P}{mgh_P}} = 2\pi\sqrt{\frac{I_C + mh_P^2}{mgh_P}}$$

where I_0, I_C and I_P are respective moments of inertia.

By substitution one easily verifies that $T_0 = T_P$ for two arrangements :

(1) The trivial solution $h_0 = h_P$ and

(2) When $I_C = mh_0h_P$.

For the second case :

$$T_0 = T_P = 2\pi\sqrt{(h_0 + h_P)/g} = 2\pi\sqrt{L/g} \quad ...(1)$$

21

Charging and Discharging of Condenser through Resistor

EXPERIMENT

OBJECT

To study the charging and discharging of a condenser through a resistor by plotting a graph of voltage (V) and current (I) across the capacitor Vs time (T) and to determine the time constant from this graph.

APPARATUS REQUIRED

Resistor, condensers, keys, D.C. power supply, milliammeter, voltmeter, potential divider, stop watch or mega type ETB-82.

FORMULA USED

Time constant T = J/0.693 where *j* is lifetime,

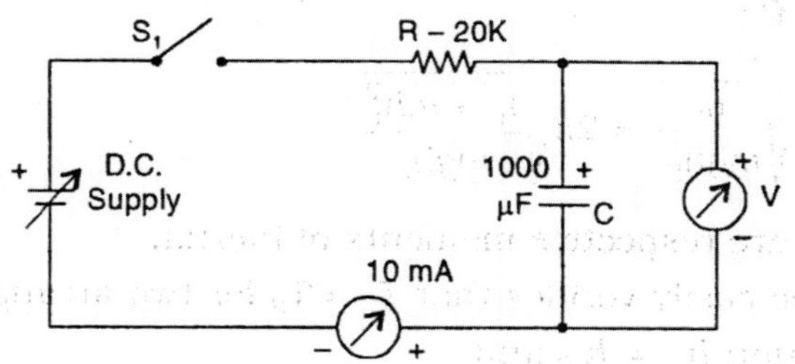

Fig. 21.1(*a*). Circuit diagram for charging of a condenser.

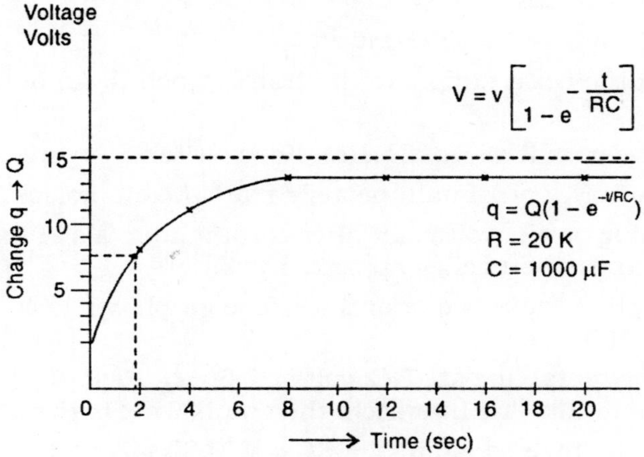

Fig. 21.1(*b*). Graph of changing voltage (V) against time (*t*).

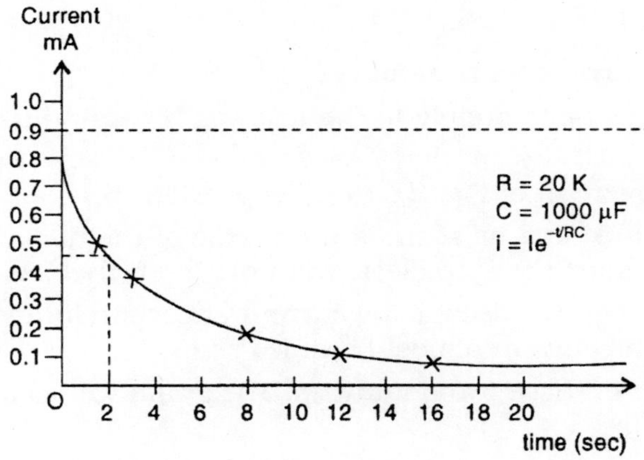

Fig. 21.1(*c*). Graph of charging current (*i*) against time (*t*).

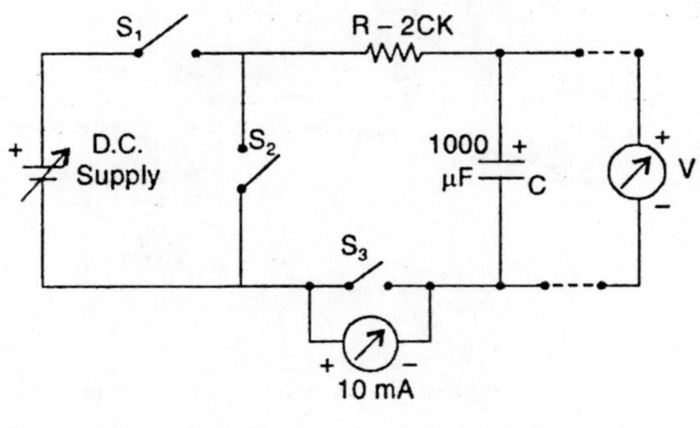

Fig. 21.2(*a*). Circuit diagram for discharge of a condenser.

EXPERIMENTAL PROCEDURE

(A) For charging voltage Vs time curve

1. Switch on the main supply to the training board. Adjust D.C. supply to 15 Volt.
2. Link up the circuit of Fig. 21.1(*a*). Keep switch S_1 open.
3. Close switch S_1 and simultaneously start the stop clock.
4. Take reading on the voltmeter after certain intervals of time (say 1 or 2 sec). till the voltage approaches approx. 15 volt.
5. Plot a graph of these two quantities. The graph will look like the one shown in Fig. 21.1(*b*).
6. Draw a horizontal line at 15/2 volt or 7.5 volt. Read the value of time on the X-axis where this line intersects the graph. This is the : life time : J.
7. Calculate the time constant T = RC = J/0.693.
8. Compare it with the theoretical time constant

i.e., $T = RC = 1000 \text{ ohm} \times 1000\ \mu F$

$10^3 \times 10^{-3} = 1$ sec.

(B) For charging current Vs time curve

1. Switch on the main supply to the training broad. Adjust D.C. supply to 15 Volt.
2. Link up the circuit of Fig. 21.1(*a*). Keep switch S_1 open.
3. Close switch S_1 and as soon as the needle of the milliammeter reaches the peak value, start the stop clock and note down the peak value of current.
4. Take reading on the decreasing current at certain intervals of time (say 1 or 2 sec.) till the current almost becomes zero.
5. Plot a graph of these two quantities. The graph will look like the one shown in Fig. 21.1(*c*).
6. Draw a horizontal line at 1/2 position (I = max. value of current). Read the value of time on the X-axis where this line intersects the graph. This is the : lifetime : J.
7. Calculate the time constant T = J/0.693.

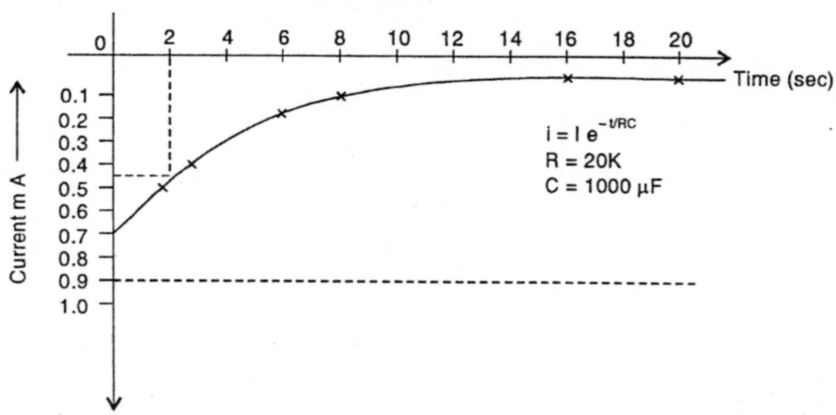

Fig. 21.2(*b*). Graph of Discharging current (*i*) Against time (*t*).

(C) For discharging voltage Vs time curve

1. Switch on the main supply to the training board. Adjust D.C. supply to 15 Volt.
2. Link up the circuit Fig. 21.2(*a*). Keep switch S_1 and S_2 open and switch S_3 closed.
3. Close switch S_1 and keep it closed till the voltage across the condenser reaches about 15 volt.
4. Now open switch S_1. Close switch S_2 and simultaneously start the stop clock. Take reading on the voltage across the condenser at certain intervals of time (say 1 or 2 sec.) till the voltage approaches approx, zero.
5. Plot a graph of these two quantities. The graph will look like the one shown in Fig. 21.2(*c*).
6. Draw a horizontal line at 15/2 volt = 7.5 volt. Read the value of time on the X-axis where this line intersects the graph. This is the : lifetime : J.
7. Calculate the time constant T = RC = J/0.693.
8. Compare it with the theoretical time constant

i.e., $T = RC = 1000 \text{ ohm} \times 1000\ \mu F$

$= 10^3 \times 10^{-3} = 1 \text{ sec.}$

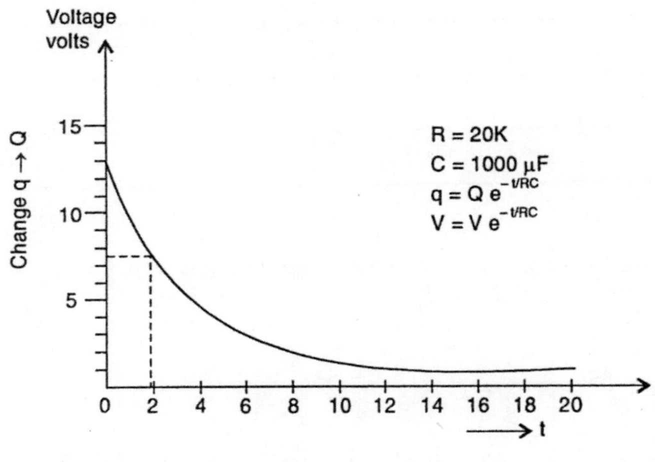

Fig. 21.2(*c*) Graph of Discharging voltage (V) Against time (*t*).

(D) For discharging current Vs time curve

1. Switch ON the main supply to the training board. Adjust D.C. supply to 15 Volt.
2. Link up the circuit of Fig. 21.2(*a*). Keep switch S_1 and S_2 open and switch S_3 closed.
3. Close switch S_1 and keep it closed till the voltage across the capacitor reaches about 15 volt.
4. Now open switch S_1 and S_3. Close switch S_2 and simultaneously start the stop clock.
5. Take reading of discharge current at certain intervals of time (say 1 or 2 sec).
6. Plot a graph of these two quantities. The graph will look like the one shown in Fig. 21.2(*b*).

7. Draw a horizontal line at 1/2 position on the graph (I = max. value of current). Read the value of time on the X-axis where this line intersects the graph. This is the : lifetime : J.
8. Calculate the time constant T = J/0.693.

OBSERVATIONS

Make the following observation table

Time (sec)	*Charging voltage (V)*

Time (sec)	*Charging current (mA)*

RESULT

Charging voltage, charging current, discharging voltage and discharging current *Vs* time curves are shown in the graph.

22

Searl's Apparatus

EXPERIMENT

OBJECT

To determine values of Y, η and σ for the material of a wire by Searl's method.

APPARATUS REQUIRED

Searl's apparatus, screwgauge, thread and stopwatch.

FORMULA USED

$$Y = \frac{8\pi K l}{T_1^2 . r^2}$$

$$\eta = \frac{8\pi K l}{T_1^2 . r^4}$$

$$\sigma = \frac{T_2^2}{2T_1^2} - 1$$

Where Y, η, σ, are Young's modulus, Modulus of rigidity and Poisson ratio respectively. K is the moment of inertia of the inertia bars.

For the square roots of mass M, length L, breadth *b*, we have

$$K = \frac{M}{12}(L^2 + b^2)$$

Whereas for cylindrical bars of mass M, length L, and radius R

$$K = M\left[\frac{L^2}{12} + \frac{R^2}{4}\right].$$

EXPERIMENTAL PROCEDURE

Suspend the inertia bars with the help of threads and adjust the length of the threads such that the bars remain parallel. Ensure that they remain horizontal as well. Clamp the experimental wire between middle points O and O′ of the bars. Tie the two ends of bars B and D with the help of a thin cotton thread and adjust its length so that these ends come closer. Allow the system to attain condition of rest. Now burn the cotton thread joining the ends B and D, the bar starts oscillating. Determine time period of the oscillation T_1 with help of a stopwatch by noting time taken for at least 20 oscillations. Repeat this process and calculate the mean value of T_1. Now take out the inertia bars and suspend either of the bars by clamping the other bar by a horizontal rigid support as shown in the Fig. 22.2. Twist the lower bar by a small angle and release it. The bar starts torsional oscillation. Determine the time period T_2 with help of a stopwatch. Repeat this observation and finally evaluate the mean value of T_2.

Measure the diameter of the wire with the help of a screwgauge into mutually perpendicular directions. Take several observations for diameter as the radius of the wire occurs in its fourth power *i.e.*, r^4. Do not forget to take observations for the zero error of the screwgauge and to correct the observation accordingly. Measure the length L of the wire OO′ and the length of the inertia bars with the help of a metre scale. Measure either the breadth *b* of the inertia bars or its radius R using a Vernier Callipers depending upon whether its cross-sectiontional is square or circle. Finally, weigh the bars using physical balance and determine their average mass.

OBSERVATIONS

1. Determination of T_1 and T_2

   Least count of stopwatch is = ... sec.

S.No.	*Number of oscillations*	*Determination of T_1*			*Determination of T_2*		
		Total time taken		*T_1*	*Total time taken*		*T_2*
		min	*sec*	*sec*	*min*	*sec*	*sec*
1.							
2.							
3.							
4.							
5.							
		Mean			Mean		

2. Observations regarding the wire

   (*i*) Length of the wire l = ... cm.

   (*ii*) L.C. of the screwgauge = ... cm.

   Zero error of the screwgauge = ± ...... cm.

S.No.	Reading along diameter A (cms)	Reading along ⊥ to diameter B (cms)	Diameter $(A + B)/2$ (cms)	Mean Diameter (cms)	Corrected mean diameter (cms)	Radius R (cms)
1.						
2.						
3.						
4.						
5.						

3. Observations regarding the bars

(*i*) Mass of one of the bars = ... gms.

Mass of the other bar = ... gms.

(*ii*) Length of one of the bars = ... cms.

Length of other bar = ... cms.

(*iii*) Breadth of one of the bar = ...cm ...cm ...cm

Breadth of the other bar = ...cm ...cm ...cm

Or

Radius of one the bars = ...cm ..,cm ....cm

Radius if other bar = ...cm ...cm ...cm.

CALCULATION

T_1 = ... sec T_2 = ... sec

L = ... cms r = ... cm.

Average mass of the bars M = ... gms

Average length of the bars L = ... cms

Average breadth of the bars b = ... cms

Average radius of the bars R = ... cms

$$K = \begin{cases} \frac{M}{12}(L^2 + b^2) & \text{for square cross section} \\ M\left(\frac{L^2}{12} + \frac{R^2}{4}\right) & \text{for circular cross section} \end{cases}$$

$$Y = \frac{8\pi Kl}{T_1^2 \cdot r^4} = \ldots\ldots \text{ dyne/cm}^2$$

$$\eta = \frac{8\pi Kl}{T_2^2 \cdot r^4} = \ldots\ldots \text{ dyne/cm}^2$$

$$\sigma = \frac{T_2^2}{2T_1^2} - 1 = \ldots\ldots$$

Probable error

1. For Y : $Y = \dfrac{8\pi K l}{T_1^2 \cdot r^4}$ ...(1)

Take logarithm of the both sides of this formula and differentiate it to get the expression

$$\frac{dY}{Y} = \frac{dl}{l} + \frac{dK}{K} + \frac{2d\,T_1}{T_1} + \frac{4Dd}{D} \quad \ldots(2)$$

Where D is the diameter of the wire = $2r$

Further

$$K = \frac{M}{12}(L^2 + b^2) \qquad \Rightarrow \qquad \frac{dK}{K} = \frac{dM}{M} + \frac{2LdL + 2bdb}{(L^2 + b^2)} \quad \ldots(3)$$

$$K = M\left(\frac{L^2}{12} + \frac{R^2}{4}\right) \qquad \Rightarrow \qquad \frac{dK}{K} = \frac{dM}{M} + \frac{2LdL + 6RdR}{(L^2 + 3R^2)} \quad \ldots(4)$$

dM = sensitivity of the physical balance and other differentials have their usual interpretations. Using these formulae calculate Y.

Note that

$$d\,T_1 = \frac{dt}{N} = \frac{\text{least count of stopwatch}}{\text{no. of oscillations}} \quad \ldots(5)$$

2. For η : $\eta = \dfrac{8\pi K l}{T_2^2 \cdot r^4}$ ...(6)

This formula is identical to one Y with the only difference of T_2 in place of T. Hence the formula (2) for the probable error applies equally well to this case with relevant modification.

3. For σ : $\sigma = \dfrac{T_2^2}{2T_1^2} - 1 \qquad \Rightarrow \qquad \sigma + 1 = \dfrac{T_2^2}{2T_1^2}$ ...(7)

Using standard method

$$\frac{d\sigma}{\sigma + 1} = \frac{2T_2}{T_2} + \frac{2dT_1}{T_1} \quad \ldots(8)$$

And $d\,T_1 = d\,T_2 = \dfrac{dt}{n}$ (see eq. (5) above) ...(9)

RESULT

The Young's modulus for the material of the wire

$Y = \ldots\ldots$ dyne/cm^2

The Modulus of rigidity of the material of the wire

$\eta = \ldots\ldots.$ dyne/cm^2

The Poisson ratio of the material of the wire

$\sigma = \ldots\ldots$

SOURCES OF ERROR AND PRECAUTIONS

1. The oscillation of the bars should be in horizontal plane. It should be ensured by adjusting the length of the threads and adjusting the knots.
2. The radius of the wire occurs at its fourth power in the formula of Y and η, hence should be determined with utmost care and accuracy.
3. Never forget to account for the zero error of the vernier calipers and screw gauge.
4. The backlash error arising during measurements by screw gauge should be avoided. This error occurs when you suddently reverse the motion of the head of the screw. To avoid this take the observations by giving motion of the head in one direction throughout a particular observation.

VIVA-VOCE

Q.1. What do you understand by elasticity ?

Ans. This is the property of the material of a body by virtue of which it (*i*) opposes a change in its size and shape arising due to some external deforming forces and (*ii*) tries to regain its original shape and size as and when the deforming forces are withdrawn.

Q.2. What do you understand by modulus of elasticity ?

Ans. Within the elastic limit the stress produced in the body because of the application of some deforming fore is proportional to the strain produced in the body (hook's law). The constant of proportionality is called the modulus of elasticity.

Q.3. How many modulus of elasticity do you encounter ?

Ans. There are three types of moduli of elasticity depending upon the nature of the deformations produced. The moduli of elasticity are called (*i*) young's modlus Y ; (*ii*) bulk modulus K ; (*iii*) shear modulus η ; depending upon whether there has been a change in its length, or volume or shape.

Q.4. What do you understand by stress ?

Ans. When a change either in the shape or in the size of the body starts place due to the application of external deforming forces there develops an opposite force which is proportional to the surface area to which the deforming force is applied. This force increases as the strain increases and ultimately when the restoring force equals the external force applied, the system attains equilibrium. The restoring force per unit area is called the stress.

Q.5. What is the principle for determination of Y ?

Ans. The basic principle for the determination of Y is that when a wire is bent into a circular arc of radius ρ there develops a restoring couple equal to $\frac{Yl}{\rho}$ where Y is the young's modulus of the material of the wire and 'l' is geometrical moment of intertia of the wire.

Q.6. What is the principle on which you determine η ?

Ans. The basic principle for the determination of η is that of the torsion of a cylinder or wire.

Q.7. How do you determine Poisson ratio ?

Ans. The Poisson ratio σ is related to young's modulus Y and modulus of rigidity η is via $\sigma = \left(\frac{Y}{2\eta - 1}\right)$. Once Y and η are known σ is calculated.

Q.8. What are theoretical limits for σ ?

Ans. The theoretical value of σ must lie between 0.5 and – 1 *i.e.*, $1 < \sigma < 0.5$.

23

Melde's Experiment

EXPERIMENT

OBJECT

To determine frequency of an electrically maintained tuning fork by Melde's method.

APPARATUS REQUIRED

Electrically maintained tuning fork, wieght box, meter scale, thread, pan and a chemical balance.

FORMULA USED

The frequency of the tuning fork is given by

In Transverse arrangement

$$n = \frac{1}{2L}\sqrt{\left(\frac{T}{M}\right)}$$

In Longitudinal arrangement

$$n = \frac{1}{L}\sqrt{\left(\frac{T}{M}\right)}$$

Where l = length of a single loop

T = Tension applied

= Mg

m = mass per unit length of the thread.

EXPERIMENTAL PROCEDURE

(A) Transverse Arrangement

Adjust electrically maintained tuning fork and thread about 3-4 metres long, along with pan according to the figure below :

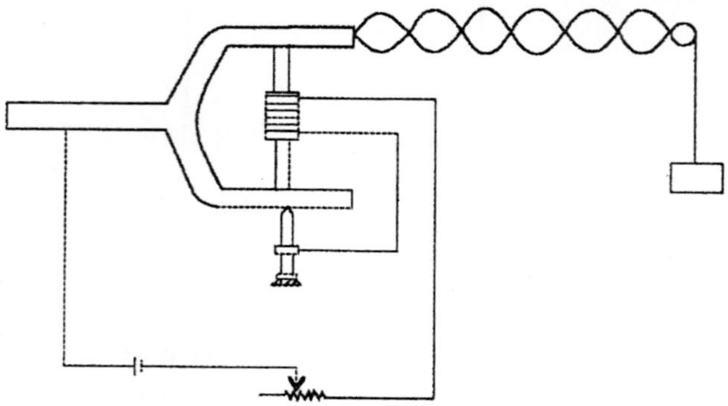

Fig. 23.1.

Put a particular mass *e.g.* 50 gm on the pan and set adjustable screw in touch with prong of tuning fork. Tuning fork starts vibration perpendicular to the vibration of the string in traverse mode. Count the number of loop on the string.

Again keep another mass like 100 gm on the pan and repeat the same process as above. Enter the data in the following observation table.

(B) For Longitudinal Arrangement

Adjust the tuning fork and the thread according to the figure given below :

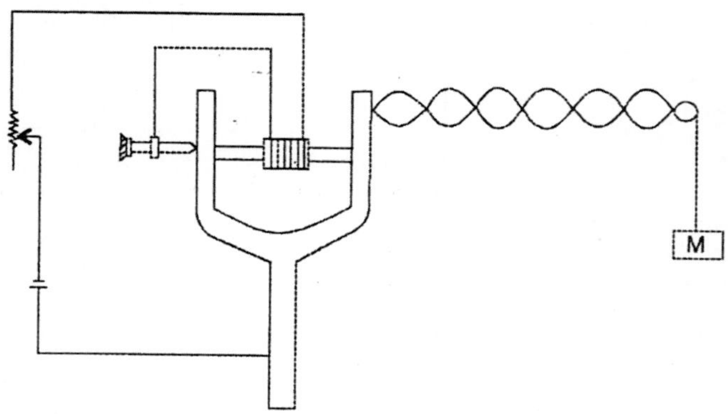

Fig. 23.2.

OBSERVATIONS

1. Transverse Arragement

S.No.	Mass kept on the pan (m1)	No. of loops	Length of the thread (cm)	Length of the thread for a single loop (cm)	Total mass applied $M = m_1 + m_2$ (gm)	$T = Mg$	$\frac{\sqrt{T}}{L}$
1.							
2.							
3.							

2. Longitudinal Arrangement

S.No.	Mass kept on the pan (m1)	No. of loops	Length of the thread (cm)	Length of the thread for a single loop (cm)	Total mass applied $M = m_1 + m_2$ (gm)	$T = Mg$	$\frac{\sqrt{T}}{L}$
1.							
2.							
3.							

3. Determination of (m).

Mass of the thread $m' =$ .... gm

Length of the thread $l' =$ .... cm

$$m = \frac{m'}{l'} \text{ gm/cm.}$$

CALCULATION

Transverse Arrangement

$$n = \frac{1}{2L}\sqrt{\left(\frac{T}{m}\right)} = \frac{1}{2\sqrt{m}}\frac{\sqrt{T}}{L}$$

= .... Hertz

Longitudinal Arrangement

$$n = \frac{1}{\sqrt{m}}\frac{\sqrt{T}}{L}$$

= .... Hertz

Mean value of frequency $n =$ .... Hz.

Maximum probable error

$$n = \frac{1}{2L}\sqrt{\left(\frac{Mg}{m}\right)}$$

$$n = \frac{1}{2L}\sqrt{\left(\frac{Mgl'}{m}\right)}$$

$$\log n = \frac{1}{2}\log M + \frac{1}{2}\log g + \frac{1}{2}\log l' - \log l - \frac{1}{2}\log m'$$

$$\frac{dn}{n} = \frac{1}{2}\frac{dM}{M} + \frac{dg}{g} + \frac{1}{2}\frac{dl'}{l} + \frac{dl'}{l} + \frac{1}{2}\frac{dm'}{m'}$$

$$\frac{dn}{n} = \frac{1}{2}\frac{dl'}{l} + \frac{dl}{l} + \frac{1}{2}\frac{dm'}{m'} \qquad (dM + dg = 0)$$

$dl = dl'$ = least count of metre scale.

dm' = minimum mass which can be measured by balance.

RESULT

The frequency of tuning fork is = $(n \pm dn)$ Hertz.

SOURCES OF ERROR AND PRECAUTIONS

1. The pulley should be frictionless.
2. There should not be any knot in the thread.
3. For getting the correct value of 'l' it is important to form well defined loops either by changing the length of the thread or by changing the mass.
4. In longitudinal arrangement it will be better to use smaller mass for providing tension. By doing so a reasonably good number of well defined loops will be obtained.

VIVA-VOCE

Q.1. What is an electrically maintained tuning fork ?

Ans. When a tuning fork is struck on a rubber pad, it starts vibrating with its natural frequency in air. The medium of air imposses resistance upon vibration and causes to dissipate the energy which ultimately stops the vibration of tuning fork. In an electrically maintained tuning fork the prongs are made to vibrate regularly by attaching one of the prongs to an electrical circuit and placing electromagnet in between the prongs.

Q.2. Explain the working of an electrically maintained tuning fork.

Ans. When adjustable screw is made to switch on the electrical circuit, the electromagnet placed in between the prongs of the tuning fork attracts the prongs and hence electrical circuit is off. The prongs start vibration and achieve their original position. As soon as prongs get their original position, adjustable screw touches the prongs and consequently electrical circuit is further switched on and thus oscillation of the prongs continues.

Q.3. What is the principle which implicates in Melde's experiment ?

Ans. Principle of stationary wave.

Q.4. How are stationary waves formed in your experiment ?

Ans. The one end is attached to the tuning fork and other is passing over a pulley and hence this string is equivalent to a string having its ends fixed. The waves originating at one end are reflected from the other end and superposition of these waves result into what is known as stationary waves.

Q.5. Explain the transverse and longitudinal mode of vibration of tuning fork in Melde's experiment and also explain why

In transverse mode

$$n = \frac{l}{2l}\sqrt{\left(\frac{T}{M}\right)}$$

In longitudinal mode

$$n = \frac{1}{l}\sqrt{\left(\frac{T}{M}\right)}.$$

Ans. If vibration of tuning fork and vibration of string are adjusted perpendicular to each other then it is said to be in transverse mode of vibration. For transverse arrangement prongs and string are taken along a line. In longitudinal arrangement vibration of tuning fork and string are adjusted parallel to each other. For longitudinal mode vibrating arms of tuning fork is perpendicular to the string.

In transverse mode

$$n = \frac{1}{2l}\sqrt{\left(\frac{T}{M}\right)}$$

Whereas is longitudinal mode

$$n = \frac{1}{l}\sqrt{\left(\frac{T}{M}\right)}$$

Because in longitudinal mode string vibrates with half of the frequency of vibration of tuning fork.

Q.6. Why is the frequency of the vibration of string in longitudinal arrangement is half of the frequency of tuning fork ?

Ans. In longitudinal arrangement the sequence of vibration appears in such a way that string completes only half of its oscillation corresponding to one complete oscillation of prong of tuning fork.

24

Plancks's Constant and Work Function

EXPERIMENT

OBJECT

To determine value of Planck's constant and the work function of a metal with the help of a photocell.

APPARATUS REQUIRED

A photocell, lamp and scale arrangement ballistic galvanometer, an optical bench, a housing with an opening for holding, a mercury vapour lamp fitted in a housing with an opening, a potentiometer, a battery or power supply, a cell, a cadmium cell or any other standard cell, three rheostats, a galvanometer, two one way key, two tapping keys and one two way key.

FORMULA USED

The Planck's constant and work function are given respectively by

$$h = \frac{eV}{\nu - \nu_0}$$

$$\phi = h\nu_0$$

e = charge of the electron,

V = stopping potential,

ν = frequency of incident radiation,

ν_0 = threshold frequency.

Basic Principle Involved. When a photon of energy *hn* is incident on a metal plate, then it is completely absorbed and in disappearing, it imparts its total energy to a single electron. A part of its total energy(f) is utilised in making the electron free from the atom and then from the metal surface and the rest is used in imparting Kinetic energy $\frac{1}{2} mv^2$ to the electron. The energy f is known as the work function. According to principle of conservation of energy

$$h\nu = \phi + \frac{1}{2} mv^2 \qquad ...(1)$$

Where ν represents frequency of incident photon m and v represents mass and velocity respectively. For a given metal ϕ is constant so, kinetic energy varies linearly with the frequency of incident photon. It does not depend upon the intensity of the incident radiation. As the intensity of the incident radiation increases the number of electrons emitted also increases.

Eqn. (1) may be used to determine h provides ν, ϕ and v are used (known). These quantities can be determined conventially by using a photocell, a ballistic galvanometer and a potentiometer connected in a circuit diagram of Fig. 24.1. The principle of determination is as discussed below.

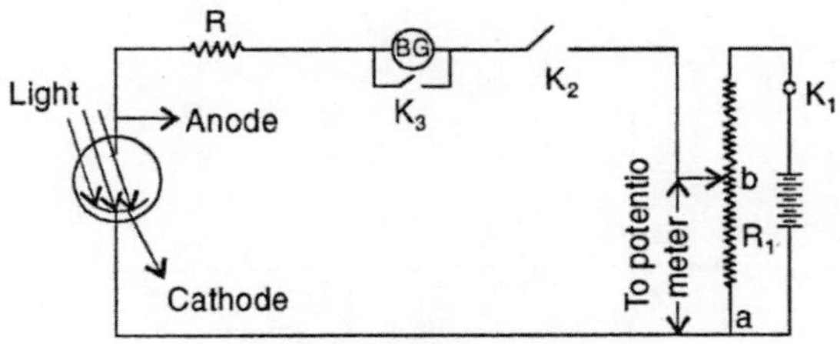

Fig. 24.1.

Determination of Work Function and the Kinetic Energy of the Electron

Let an electromagnetic radiation of known frequency ν be allowed to fall on the cathode of the photocell, the kinetic energy of the electrons emitted may not be sufficient to reach the anode. However, if we apply a +ve potential at the anode a photo current will be observed to flow through the ballistic galvanometer placed in the circuit. In case we keep the anode at the –ve potential the photo electrons will be repelled by the anode and consequently the photo current will decrease. Now, if the anode is made more –ve by increasing its potential there comes a stage at which the photo electrons seases to flow and the photo current in this condition will comes to O. This potential is known as the stopping potential. Let this potential be V. Under the situation, when the stopping potential is applied between the cathode and the anode of the photocell, the electrostatic potential energy due to the repulsive force acting on the electrons is equal to their kinetic energy. Hence,

$$eV = \frac{1}{2} mv^2 \qquad ...(2)$$

where e = electronic charge.

In the light of eqn. (2), eqn. (1) transforms as

$$h\nu = \phi + eV \qquad ...(3)$$

Eqn. (3) suggests that from different value of ν, V have different values of stopping potential (V). In other words, on using radiations different wavelengths of different values of stopping potentials will be obtained. If we plot a graph between V and the frequency of incident radiation ν by taking the former on Y-axis we get a straight line as shown in Fig. 24.2. This line cuts the X-axis at the point B where V = 0. Let the corresponding frequency be ν_0 then Eqn. (3) gives

$$\boxed{h\nu_0 = \phi} \quad ...(4)$$

The frequency ν_0 represents the minimum value of the frequency of incident photon just sufficient to free an e^- from metal surface and called *Threshold frequency*. The slope of the straight line is

$$\boxed{\tan\theta = \frac{h}{e}} \quad ...(5)$$

Using Eqn. (5) value of Planck's constant can be determined. Once h is known, we can determine the value of work function of metal of the cathode by using Eqn. (4).

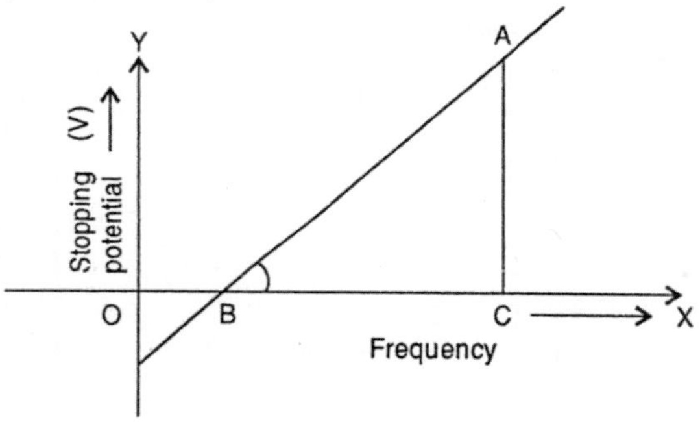

Fig. 24.2.

EXPERIMENTAL PROCEDURE

Using a battery or power supply construct a potential devider and complete the electrical connections as shown in Fig. 24.1. Switch on the source of light of the lamp and scale arrangement and allow the light to fall on the mirror of the galvanometer and form a bright image of the cross wire. Coinside the cross wire with the O of the scale before the start of the experiment, it is necessary to check whether the galvanometer is oscillating freely or not. In the case the motion is hindered then get rid of it by leveling the base of the galvanometer. Fix a photocell in its housing and mount it at one end of the optical bench. Similarly fix a mercury vapour lamp in its housing and fix it at the other end of the optical bench. The distance between the source and the photocell should be kept fixed throughout the procedure of experiment. It is important to note that the opening of source of light should face the opening of photocell housing.

Insert a filter which can isolate a 4046Å radiation of mercury, in a socket built in the housing of the source. Allow the transmitted light from the filter to fall on the photocell. With the help of potential devider set the anode to some –ve potential. Press the tapping key K_2 and note the steady deflection of the scale. Adjust the anode potential with the help of rheostat R_1 to the –ve value such that on pressing the key K_2 no deflection is observed. This potential is the stopping potential at 4046Å. The measurement of this potential should be very accurate and it should be done with the help of potentiometer.

Using different filters find out the stopping potential exactly in the similar way, at wavelengths 4358Å, 5461Å, 5770Å and 5790Å of mercury. Draw a graph *b/w* the frequency of incident radiation and stopping potential with frequency along X-axis and V along Y-axis. A straight line identical to Fig. 24.2 is obtained. Determine *h* and by knowing slope of straight line and intercept OB.

Measurement of Potential

Standardise the potentiometer wire. Connect the point marked *a* and *b* of the circuit diagram of Fig. 24.1 to the unknown gap of potentiometer following the procedure given under the heading collaboration of voltmeter, determine the potential difference V across the cathode and anode of the photocell.

OBSERVATIONS

(*a*) Data :

$$e = 1.6 \times 10^{-19}\ \text{C}$$

$$1.6 \times 10^{-19}\ \text{Joule} = 1\ \text{eV}$$

$$c = 3 \times 10^{8}\ \text{m/sec.}$$

(*b*) Determine the stopping potential

S.No.	Wavelength obtained using filters λ, Å	Frequency corresponding to the wavelength λ. $\nu = c/\lambda\ \text{sec}^{-1}$	Stopping potential measured by potentiometer (V) volt
1.	4046		
2.	4358		
3.	5461		
4.	5770		
5.	5790		

CALCULATION

From graph V-ν

AC = ... cm

BC = ... cm

Slope of the straight line $\frac{h}{e} = \frac{AC}{BC}$ ......

h = ... Joule-sec.

OB = ν_0 = ... vibrations/sec.

$$\phi = h\,.\,OB = \frac{AC}{BC}\,.\,OB\,.\,e$$

= ...... Joule = ... eV

PROBABLE ERROR

$$h = \frac{e\,\text{AC}}{\text{BC}} = \frac{e\,\text{AC}}{\text{OC} - \text{OB}}$$

taking logarithm of both the sides

log h = log AC – log (OC – OB)

differentiating, we get and taking +ve sign

$$\frac{dh}{h} = \frac{d(\text{AC})}{\text{AC}} + \frac{d(\text{OC}) + d(\text{OB})}{(\text{OC} - \text{OB})}$$

$$\phi = \text{OB} \,.\, \frac{\text{AC}}{\text{BC}} \times e = \frac{e\,\text{OB}\,.\,\text{AC}}{(\text{OC} - \text{OB})}$$

Using the standard procedure, we get

$$\frac{d\phi}{\phi} = \frac{d(\text{OB})}{\text{OB}} + \frac{d(\text{AC})}{\text{AC}} + \frac{d(\text{OC}) + d(\text{OB})}{(\text{OC} - \text{OB})}$$

d(AC) = Leastcount of the graph along Y-axis,

d(OC) = d(OB) = Leastcount of the graph along X-axis.

On substituting values, we get the value of dh and $d\phi$.

RESULT

(*a*) The value of Planck's constant = ... J. sec,

(*b*) The work function of metal cathode = ... eV.

SOURCES OF ERROR AND PRECAUTIONS

1. The galvanometer used should be sensitive.
2. One should not use voltmeter for determining the stopping potential. It should be done by potentiometer.
3. A good quality filter should be used.
4. Throughout the experiment the distance between the source and photocell should be kept constant.
5. No stray radiation should be allowed to fall on the photocell during experiment.

VIVA-VOCE

Q.1. What is Photo electric effect ?

Ans. Emission of e^- from metal surface due to incidence of light of suitable frequency.

Q.2. What is stopping potential ?

Ans. The potential at which the e^- just seases to flow is called stopping potential.

Q.3. What is threshold frequency ?

Ans. The minimum frequency at which the e^- eject from the metal surface with zero kinetic energy.

Q.4. Define work function.

Ans. Work function is the amount of energy to just release the e^- from the metal surface.

Q.5. Why mercury vapour lamp is used in the experiment ?

Ans. Because it is not monochromatic source of light.

25

Diameter of a Thin Wire

EXPERIMENT

OBJECT

To determine the diameter of a thin wire by interference due to a wedge shape air film.

APPARATUS REQUIRED

Bridge type microscope, sodium vapour lamp, Newton's ring assembly and the wedge formed by using the given wire.

FORMULA USED

$$D = \frac{\lambda l}{2\bar{x}}$$

Basic Principle Involved

Let us consider a wedge shape air film with the help of a plane glass plates OA and OB and wire W Fig. 25.1. The angle of wedge $\theta = \frac{D}{l}$.

D = The diameter of wire.

When the air film is viewed from above with reflected monochromatic light, a system of equidistant interference fringes are observed. These fringes are parallel to the line of intersection of the two surfaces. Consider the situation in which the rays obtained from the extended monochromatic source are viewed through the microscope at nearly normal incidence after reflection from the wedge. Suppose the nth bright fringe occurs at P_n Fig. 25.1. The thickness of air film at

$$P_n = P_nQ_n$$

Using the conditions for bright fringe, we have

$$2ut \cos r = (2n + 1)\frac{\lambda}{2} \qquad ...(1)$$

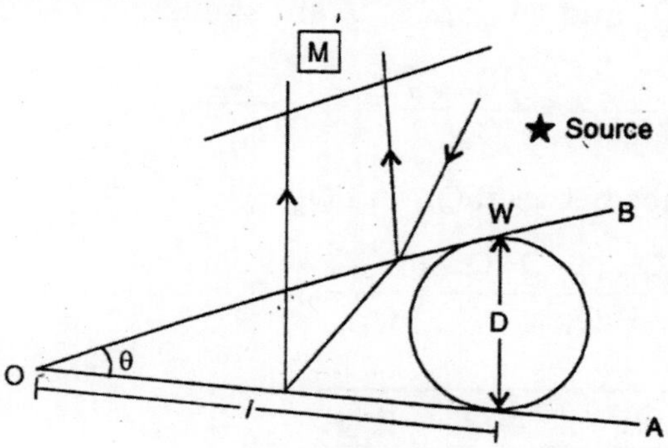

Fig. 25.1.

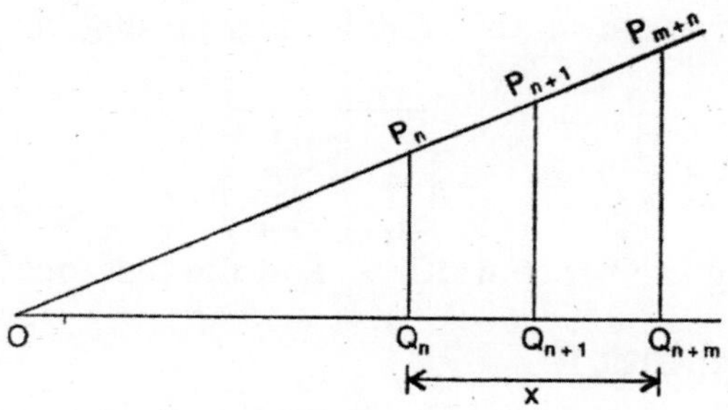

Fig. 25.2.

In Eqn. (1) is case $u = 1$, $r = 0$ and $t = P_nQ_n$

Hence,
$$2P_nQ_n = (2n + 1)\frac{\lambda}{2} \qquad ...(2)$$

For the next bright fringe, we have

$$2P_{n+1}Q_{n+1} = [2(n + 1) + 1]\frac{\lambda}{2} \qquad ...(3)$$

From equation (2) and (3), we have

$$P_{n+1}Q_{n+1} - P_nQ_n = \frac{\lambda}{2} \qquad ...(4)$$

This equation suggests that the next bright fringe will occur at the point where the thickness of the air film increases by $\frac{\lambda}{2}$.

From $(m + n)^{th}$ fringe, we have

$$\boxed{P_{m+n}Q_{m+n} - P_nQ_n = m\frac{\lambda}{2}} \quad ...(5)$$

The triangle OP_nQ_n and $OP_{n+m}Q_{n+m}$ are similar.

Hence,
$$\frac{P_{m+n}Q_{m+n}}{P_nQ_n} = \frac{OQ_{m+n}}{OQ_n} = \frac{x + OQ_n}{OQ_n}$$

where, x = distance between Q_n and Q_{m+n}

or
$$\frac{P_{m+n}Q_{m+n} - P_nQ_n}{P_nQ_n} = \frac{x}{OQ_n}$$

or
$$\boxed{\frac{P_{m+n}Q_{m+n} - P_nQ_n}{x} = \frac{P_nQ_n}{OQ_n} = \theta} \quad ...(6)$$

The expression for fringe width is obtained by putting $m = 1$ and $x = \overline{x}$ using Eqn. (5), we have

$$\frac{\lambda}{2\overline{x}} = \theta = \frac{D}{l} \quad \text{or} \quad \boxed{D = \frac{\lambda l}{2\overline{x}}} \quad ...(7)$$

Thus, by knowing the fringe width $\overline{x}$ and the distance l, D can be determined.

EXPERIMENTAL PROCEDURE

Perform the general setting of the source and glass plate G, for the Newton ring assembly. Place the wedge on a piece of white paper. Allow, the light to fall on the transparent glass plate G. Adjust the inclination of plate G in order to get a proper intensity of the reflected light. Observe the equidistant bright and dark straight line fringes through the Microscope M. Measure the position of a series of fringes and from these observation determine the fringewidth $\overline{x}$. Measure l using travelling microscope. It will be better to take a few observations for l and then find out the average value of l.

OBSERVATIONS

Leastcount of travelling microscope = ... cm

Average value of length l = ... cm

S.No.	*Position of the fringes* cm	*Seperation for 10 fringes* cm	*Average separation for 10 fringes* cm	*Fringe width* $\overline{x}$ cm
1.				
2.				
3.				
4.				
5.				
⋮				
17.				
18.				
19.				
20.				

CALCULATION

$$D = \frac{\lambda l}{2\overline{x}}$$

$$= \ldots\ldots \text{ cm}$$

Probable Error

$$D = \frac{\lambda l}{2\overline{x}} = \frac{\lambda l}{2(x_2 - x_1)/N} = \frac{\lambda /N}{2(x_2 - x_1)}$$

Here, x_2 and x_1 represents the positions of the fringes at a separation of N fringes. Taking log of this equation and then differentiating, we get

$$\log D = \log \lambda + \log l + \log N - \log 2 - \log (x_2 - x_1)$$

$$\frac{dD}{D} = \frac{d\lambda}{\lambda} + \frac{dl}{l} + \frac{dN}{N} + \frac{dx_2 + dx_1}{x_2 - x_1}$$

From the above equation determine the value of dD.

RESULT

The diameter of the wire = ... cm.

SOURCE OF ERROR AND PRECAUTIONS

1. Care should be taken regarding the measurement of the fringe width as the fringes has finite width.
. Care should be allowed to take the measure of the length of wedge.
3. The microscope should be allowed to move only in one direction in order to avoid backlash error.

VIVA-VOCE

See Experiment 2.

Appendix

Description of Spectrometer

The spectrometer consist of essentially three main parts :

(*a*) Collimator,

(*b*) Telescope,

(*c*) Prism table,

COLLIMATOR

The purpose of Collimator is to provide a parallel beam of light. This is achieved by a sleet and achromatic lens lying at the ends of Collimator. The Collimator is made up of a horizontal brass tube, to one end of which an achromatic lens is fitted. The other end of the tube contains another brass tube inside it, which can move along the main tube with the help of rack and pinion arrangement. The outer end of the inner brass tube contains a vertical slit. The slit is adjusted so that it lies in the first focal plane of achromatic lens. A screen C_1 at the bottom of the Collimator is provided to change its incarnation with the horizontal at the inner extremity of the Collimator tube to screws are provided whose relative movement can change its orientation within the horizontal plane. These screws are used to orient the Collimator, so, that the beam of light immerging out of it passes through the axis of the prism table. However, this setting is normally done by the manufacturer. The students are therefore, advised not to use these screws. The Collimator is mounted on the fixed rigid stand attached to the body of the spectrometer.

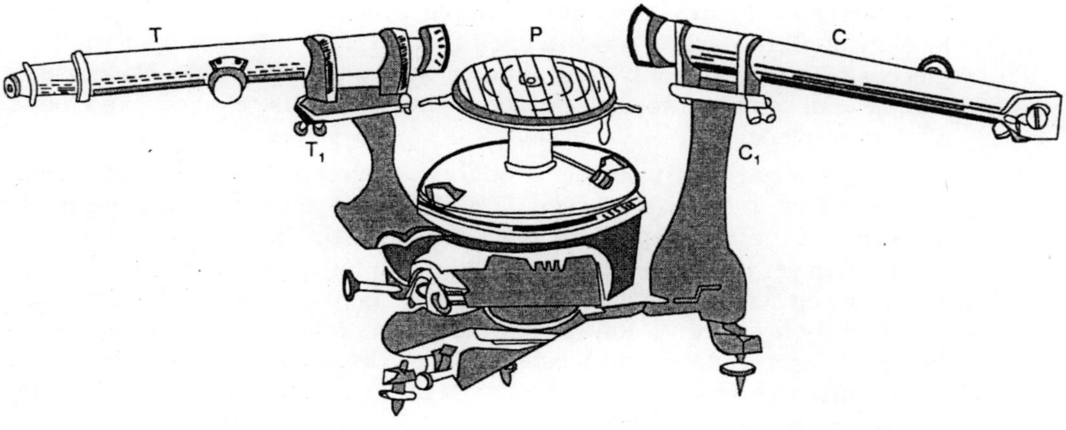

TELESCOPE

The purpose of the Telescope is to receive parallel rays. Its consist of brass tube, one end of which contains objectives made up of an achromatic lens. The outer end of the inner tube contains another tube fitted with the either Ramsden or Gauss Eyepiece along with cross wire. The inner tube as a whole is a capable of sliding forward and backward with the help of rack and pinion arrangement. The telescope is attached to a graduated metallic circular scale lying underneath the prism table. Two verniers suppurated by an angle 180° are also attached to the circular scale. In order to change the incarnation of the Telescope in the horizontal plane a screw T_1 is provided at the bottom of the Telescope, it is also provided with a clamp and a tangential screw to rotate it more accurately and slowly.

PRISM TABLE

The Prism table P is a circular plate capable of rotating about an axis of rotation. Its motion can be read with the help of two verniers suppurated by 180° apart sliding over the circular scale lying under neath the prism table. There is also provision for controlled motion of the prism table with the help of a clamp and a tangential screw. There are circular and straight grooves cut on its surface which helps in the placement of the prism on a prism table. Three screws containing screen are also provided underneath the tips of straight grooves. Making use of these the plane of the prism table can be adjusted. The prism table as a whole can be adjusted to a suitable height by way of raising and lowering it with the help of a clamp screw provided for the purpose.

SETTING OF SPECTROMETER EQUIPPED WITH RAMSDEN EYPIECE

Ramsden Eyepiece

It's consists of two plano convex lenses of equal focal length suppurated by a distance equal to two third of the focal length of either. The convex faces are towards each other. The eye piece is placed beyond the image form by the objective. It is equipped with a cross wire.

Adjustment of the Telescope and the Collimator

1. **With distant object.** Take out the eyepiece assembly from the telescope and look through it towards a white back ground. Focus it by displacing the eye piece within the assembly so that cross wire is clearly cisible. Now insert the eye piece assembly into the telescope. This setting is referred to as focusing of the eye piece.

   Look at a very distant object through the telescope. In general, a blurred image of the object will be seen. This indicates that the focal plane of the objective of the telescope does not coincide with the focal plane of the eye piece. In order to have a common focal plane for the two, move the eye piece assembly either inword or outword with the help of rack and pinion arrangements. When the condition of common focal plane will be obtain, a distinct image of distant object will be seen at the cross wire. Further, there should not be any parallax between the image and the cross wire. This setting ensures that telescope is ready to receive a beam of parallel rays.

   Tern the telescope towards the collimator and make them collinear. Illuminate the slit of the collimator with the help of a light source and adjust the width of the slit in order to get a line pensil of light emerging out of it. On looking at the slit through the telescope again a blurred image, in general, is observed. This indicates that the slit of collimator is not in the focal plane of its objectives. Using rack and pinion arrangements of the collimator displays the slit to have a distinct image of the slit on cross wire of telescope. This setting ensures that the collimator is sending a beam of a parallel rays.

2. **Schuster's method.** When a distant object is not in our reach, for example, when the experiment is being performed in a dark room. Schuster's method is used for setting the collimator and the telescope for parallel rays. Place the prism on the prism table such that one of its reflecting surface AC is perpendicular to the line joining the two screws. Switch on the source of light, say mercury vapour lamp. A uncollimated beam of light will emerge from the collimator C and fall on one of the reflecting faces AB of prism and after refraction will emerge out in the form of spectrum which can be seen from the naked eye. Now rotate the telescope to receive the spectrum. Using the rack and pinion arrangement of the telescope and focus the spectrum. Amongst the several spectral lines of different colours choose on spectral line, say green. Coincide it with the cross wire of the telescope by using the clamp and the tangential screw. On rotating the prism table the spectrum as a whole starts leaving the field of view of the telescope. Continue to rotate the prism and follow the spectrum. In the proces of rotating the prism by way of rotating the prism table and of following the spectrum through telescope you will come across a condition at which any further rotation of the prism reverses the motion of the green line of the spectrum. This condition at which the green line begins to return is known as the condition of minimum deviation for the green line of the spectrum. Turn the telescope in a position corresponding to slightly large deviation and clamp it. Now rotate the prism table by a small angle in a direction such that the angle of incidence of the beam of light increases. The spectrum in the telescope become blurred. Focus it with the help of rack and pinion arrangement of the telescope. Similarly, on slightly decreasing the angle of incidence, the spectrum will again be blurred. Focus this time with the help of rack and pinion of the collimator. The whole process is repeated till the spectrum remains sharply focussed in both the positions of prism corresponding to the increasing and decreasing the angle of incidence. The collimator and the telescope both are now set for sending and receiving the parallel beam of light respectively.

LOGARITHMIC
AND
OTHER TABLES

LOGARITHMS

	0	1	2	3	4	5	6	7	8	9	1	2	3	4	5	6	7	8	9
10	0000	0043	0086	0128	0170						5	9	13	17	21	26	30	34	38
						0212	0253	0294	0334	0374	4	8	12	16	20	24	28	32	36
11	0414	0453	0492	0531	0569						4	8	12	16	20	23	27	31	35
						0607	0645	0682	0719	0755	4	7	11	15	18	22	26	29	33
12	0792	0828	0864	0899	0934						3	7	11	14	18	21	25	28	32
						0969	1004	1038	1072	1106	3	7	10	14	17	20	24	27	31
13	1139	1173	1206	1239	1271						3	6	10	13	16	19	23	26	29
						1303	1335	1367	1399	1430	3	7	10	13	16	19	22	25	29
14	1461	1492	1523	1553	1584						3	6	9	12	15	19	22	25	28
						1614	1644	1673	1703	1732	3	6	9	12	14	17	20	23	26
15	1761	1790	1818	1847	1875						3	6	9	11	14	17	20	23	26
						1903	1931	1959	1987	2014	3	6	8	11	14	17	19	22	25
16	2041	2068	2095	2122	2148						3	6	8	11	14	16	19	22	24
						2175	2201	2227	2253	2279	3	5	8	10	13	16	18	21	23
17	2304	2330	2355	2380	2405						3	5	8	10	13	15	18	20	23
						2430	2455	2480	2504	2529	3	5	8	10	12	15	17	20	22
18	2553	2577	2601	2625	2648						2	5	7	9	12	14	17	19	21
						2672	2695	2718	2742	2765	2	4	7	9	11	14	16	18	21
19	2788	2810	2833	2856	2878						2	4	7	9	11	13	16	18	20
						2900	2923	2945	2967	2989	2	4	6	8	11	13	15	17	19
20	3010	3032	3054	3075	3096	3118	3139	3160	3181	3201	2	4	6	6	11	13	15	17	19
21	3222	3243	3263	3284	3304	3324	3345	3365	3385	3404	2	4	6	8	10	12	14	16	18
22	3424	3444	3464	3483	3502	3522	3541	3560	3579	3598	2	4	6	8	10	12	14	15	17
23	3617	3636	3655	3674	3692	3711	3729	3747	3766	3784	2	4	6	7	9	11	13	15	17
24	3802	3620	3838	3856	3874	3892	3909	3927	3945	3962	2	4	5	7	9	11	12	14	16
25	3979	3997	4014	4031	4048	4065	4082	4099	4116	4133	2	3	5	7	9	10	12	14	15
26	4150	4166	4183	4200	4216	4232	4249	4265	4281	4298	2	3	5	7	8	10	11	13	15
27	4314	4330	4346	4362	4378	4393	4409	4425	4440	4456	2	3	5	6	8	9	11	13	14
28	4472	4487	4502	4518	4533	4548	4564	4579	4594	4609	[illegible]	3	5	6	8	9	11	12	14
29	4624	4639	4654	4669	4683	4698	4713	4728	4742	4757	1	3	4	6	7	9	10	12	13
30	4771	4786	4800	4814	4829	4843	4857	4871	4886	4900	1	3	4	6	7	9	10	11	13
31	4914	4928	4942	4955	4969	4983	4997	5011	5024	5038	1	3	4	6	7	8	10	11	12
32	5052	5065	5079	5092	5105	5119	5132	5145	5159	5172	1	3	4	5	7	8	9	11	12
33	5185	5198	5211	5224	5237	5250	5263	5276	5289	5302	1	3	4	5	6	8	9	10	12
34	5315	5328	5340	5353	5366	5378	5391	5403	5416	5428	1	3	4	5	6	8	9	10	11
35	5441	5453	5465	5478	5490	5502	5515	5527	5539	5551	1	2	4	5	6	7	9	10	11
36	5563	5575	5587	5599	5611	5623	5635	5647	5658	5670	1	2	4	5	6	7	8	10	11
37	5682	5694	5705	5717	5729	5740	5752	5763	5775	5786	1	2	3	5	6	7	8	9	10
38	5798	5809	5821	5832	5843	5855	5866	5877	5888	5899	1	2	3	5	6	7	8	9	10
39	5911	5922	5933	5944	5955	5966	5977	5988	5999	6010	1	2	3	4	5	7	8	9	10
40	6021	6031	6042	6053	6064	6075	6085	6096	6107	6117	1	2	3	4	5	6	8	9	10
41	6128	6138	6149	6160	6170	6180	6191	6201	6212	6222	1	2	3	4	5	6	7	8	9
42	6232	6243	6253	6263	6274	6284	6294	6304	6314	6325	1	2	3	4	5	6	7	8	9
43	6335	6345	6355	6365	6375	6385	6395	6405	6415	6425	1	2	3	4	5	6	7	8	9
44	6435	6444	6454	6464	6474	6484	6493	6503	6513	6522	1	2	3	4	5	6	7	8	9
45	6532	6542	6551	6561	6571	6580	6590	6599	6609	6618	1	2	3	4	5	6	7	8	9
46	6628	6637	6646	6656	6665	6675	6684	6693	6702	6712	1	2	3	4	5	6	7	7	8
47	6721	6730	6739	6749	6758	6767	6776	5785	6794	6803	1	2	3	4	5	5	6	7	8
48	6812	6821	6830	6839	6848	6857	6866	6875	6884	6893	1	2	3	4	4	5	6	7	8
49	6902	6911	6920	6928	6937	[illegible]	6955	6964	6972	6981	1	2	3	4	4	5	6	7	8

LOGARITHMS

	0	1	2	3	4	5	6	7	8	9	1 2 3	4 5 6	7 8 9
50	6990	6998	7007	7016	7024	7033	7042	7050	7059	7067	1 2 3	3 4 5	6 7 8
51	7076	7084	7093	7101	7110	7118	7126	7135	7143	7152	1 2 3	3 4 5	6 7 8
52	7160	7168	7177	7185	7193	7202	7210	7218	7226	7235	1 2 2	3 4 5	6 7 7
53	7243	7251	7259	7267	7275	7284	7292	7300	7308	7316	1 2 2	3 4 5	6 6 7
54	7324	7332	7340	7348	7356	7364	7372	7380	7388	7396	1 2 2	3 4 5	6 6 7
55	7404	7412	7419	7427	7435	7443	7451	7459	7466	7474	1 2 2	3 4 5	5 6 7
56	7482	7490	7497	7505	7513	7520	7528	7536	7543	7551	1 2 2	3 4 5	5 6 7
57	7559	7566	7574	7582	7589	7597	7604	7612	7619	7627	1 2 2	3 4 5	5 6 7
58	7634	7642	7649	7657	7664	7672	7679	7686	7694	7701	1 1 2	3 4 4	5 6 7
59	7709	7716	7723	7731	7738	7745	7752	7760	7767	7774	1 1 2	3 4 4	5 6 7
60	7782	7789	7796	7803	7810	7818	7825	7832	7839	7846	1 1 2	3 4 4	5 6 6
61	7853	7860	7868	7875	7882	7889	7896	7903	7910	7917	1 1 2	3 4 4	5 6 6
62	7924	7931	7938	7945	7952	7959	7966	7973	7980	7987	1 1 2	3 3 4	5 6 6
63	7993	8000	8007	8014	8021	8028	8035	8041	8048	8055	1 1 2	3 3 4	5 5 6
64	8062	8069	8075	8082	8089	8096	8102	8109	8116	8122	1 1 2	3 3 4	5 5 6
65	8129	8136	8142	8149	8156	8162	8169	8176	8182	8189	1 1 2	3 3 4	5 5 6
66	8195	8202	8209	8215	8222	8228	8235	8241	8248	8254	1 1 2	3 3 4	5 5 6
67	8261	8267	8274	8280	8287	8293	8299	8306	8312	8319	1 1 2	3 3 4	5 5 6
68	8325	8331	8338	8344	8351	8357	8363	8370	8376	8382	1 1 2	3 3 4	4 5 6
69	8388	8395	8401	8407	8414	8420	8426	8432	8439	8445	1 1 2	2 3 4	4 5 6
70	8451	8457	8463	8470	8476	8482	8488	8494	8500	8506	1 1 2	2 3 4	4 5 6
71	8513	8519	8525	8531	8537	8543	8549	8555	8561	8567	1 1 2	2 3 4	4 5 5
72	8573	8579	8585	8591	8597	8603	8609	8615	8621	8627	1 1 2	2 3 4	4 5 5
73	8633	8639	8645	8651	8657	8663	8669	8675	8681	8686	1 1 2	2 3 4	4 5 5
74	8692	8698	8704	8710	8716	8722	8727	8733	8739	8745	1 1 2	2 3 4	4 5 5
75	8751	8756	8762	8768	8774	8779	8785	8791	8797	8802	1 1 2	2 3 3	4 5 5
76	8808	8814	8820	8825	8831	8837	8842	8848	8854	8859	1 1 2	2 3 3	4 5 5
77	8865	8871	8876	8882	8887	8893	8899	8904	8910	8915	1 1 2	2 3 3	4 4 5
78	8921	8927	8932	8938	8943	8949	8954	8960	8965	8971	1 1 2	2 3 3	4 4 5
79	8976	8982	8987	8993	8998	9004	9009	9015	9020	9025	1 1 2	2 3 3	4 4 5
80	9031	9036	9042	9047	9053	9058	9063	9069	9074	9079	1 1 2	2 3 3	4 4 5
81	9085	9090	9096	9101	9106	9112	9117	9122	9128	9133	1 1 2	2 3 3	4 4 5
82	9138	9143	9149	9154	9159	9165	9170	9175	9180	9186	1 1 2	2 3 3	4 4 5
83	9191	9196	9201	9206	9212	9217	9222	9227	9232	9238	1 1 2	2 3 3	4 4 5
84	9243	9248	9253	9258	9263	9269	9274	9279	9284	9289	1 1 2	2 3 3	4 4 5
85	9294	9299	9304	9309	9315	9320	9325	9330	9335	9340	1 1 2	2 3 3	4 4 5
86	9345	9350	9355	9360	9365	9370	9375	9380	9385	9390	1 1 2	2 3 3	4 4 5
87	9395	9400	9405	9410	9415	9420	9425	9430	9435	9440	0 1 1	2 2 3	3 4 4
88	9445	9450	9455	9460	9465	9469	9474	9479	9484	9489	0 1 1	2 2 3	3 4 4
89	9494	9499	9504	9509	9513	9518	9523	9528	9533	9538	0 1 1	2 2 3	3 4 4
90	9542	9547	9552	9557	9562	9566	9571	9576	9581	9586	0 1 1	2 2 3	3 4 4
91	9590	9595	9600	9605	9609	9614	9619	9624	9628	9633	0 1 1	2 2 3	3 4 4
92	9638	9643	9647	9652	9657	9661	9666	9671	9675	9680	0 1 1	2 2 3	3 4 4
93	9685	9689	9694	9699	9703	9708	9713	9717	9722	9727	0 1 1	2 2 3	3 4 4
94	9731	9736	9741	9745	9750	9754	9759	9763	9768	9773	0 1 1	2 2 3	3 4 4
95	9777	9782	9786	9791	9795	9800	9805	9809	9814	9818	0 1 1	2 2 3	3 4 4
96	9823	9827	9832	9836	9841	9845	9850	9854	9859	9863	0 1 1	2 2 3	3 4 4
97	9868	9872	9877	9881	9886	9890	9894	9899	9903	9908	0 1 1	2 2 3	3 4 4
98	9912	9917	9921	9926	9930	9934	9939	9943	9948	9952	0 1 1	2 2 3	3 4 4
99	9956	9961	9965	9969	9974	9978	9983	9987	9991	9996	0 1 1	2 2 3	3 3 4

ANTILOGARITHMS

	0	1	2	3	4	5	6	7	8	9	1	2	3	4	5	6	7	8	9
.00	1000	1002	1005	1007	1009	1012	1014	1016	1019	1021	0	0	1	1	1	1	2	2	2
.01	1023	1026	1028	1030	1033	1035	1038	1040	1042	1045	0	0	1	1	1	1	2	2	2
.02	1047	1050	1052	1054	1057	1059	1062	1064	1067	1069	0	0	1	1	1	1	2	2	2
.03	1072	1074	1076	1079	1081	1084	1086	1089	1091	1094	0	0	1	1	1	1	2	2	2
.04	1096	1099	1102	1104	1107	1109	1112	1114	1117	1119	0	1	1	1	1	2	2	2	2
.05	1122	1125	1127	1130	1132	1135	1138	1140	1143	1146	0	1	1	1	1	2	2	2	2
.06	1148	1151	1153	1156	1159	1161	1164	1167	1169	1172	0	1	1	1	1	2	2	2	2
.07	1175	1178	1180	1183	1186	1189	1191	1194	1197	1199	0	1	1	1	1	2	2	2	2
.08	1202	1205	1208	1211	1213	1216	1219	1222	1225	1227	0	1	1	1	1	2	2	2	3
.09	1230	1233	1236	1239	1242	1245	1247	1250	1253	1256	0	1	1	1	1	2	2	2	3
.10	1259	1262	1265	1268	1271	1274	1276	1279	1282	1285	0	1	1	1	1	2	2	2	3
.11	1288	1291	1294	1297	1300	1303	1306	1309	1312	1315	0	1	1	1	2	2	2	2	3
.12	1318	1321	1324	1327	1330	1334	1337	1340	1343	1346	0	1	1	1	2	2	2	2	3
.13	1349	1352	1355	1358	1361	1365	1368	1371	1374	1377	0	1	1	1	2	2	2	3	3
.14	1380	1384	1387	1390	1393	1396	1400	1403	1406	1409	0	1	1	1	2	2	2	3	3
.15	1413	1416	1419	1422	1426	1429	1432	1435	1439	1442	0	1	1	1	2	2	2	3	3
.16	1445	1449	1452	1455	1459	1462	1466	1469	1472	1476	0	1	1	1	2	2	2	3	3
.17	1479	1483	1486	1489	1493	1496	1500	1503	1507	1510	0	1	1	1	2	2	2	3	3
.18	1514	1517	1521	1524	1528	1531	1535	1538	1542	1545	0	1	1	1	2	2	2	3	3
.19	1549	1552	1556	1560	1563	1567	1570	1574	1578	1581	0	1	1	1	2	2	3	3	3
.20	1585	1589	1592	1596	1600	1603	1607	1611	1614	1618	0	1	1	1	2	2	3	3	3
.21	1622	1626	1629	1633	1637	1641	1644	1648	1652	1656	0	1	1	2	2	2	3	3	3
.22	1660	1663	1667	1671	1675	1679	1683	1687	1690	1694	0	1	1	2	2	2	3	3	3
.23	1698	1702	1706	1710	1714	1718	1722	1726	1730	1734	0	1	1	2	2	2	3	3	4
.24	1738	1742	1746	1750	1754	1758	1762	1766	1770	1774	0	1	1	2	2	2	3	3	4
.25	1778	1782	1786	1791	1795	1799	1803	1807	1811	1816	0	1	1	2	2	2	3	3	4
.26	1820	1824	1828	1832	1837	1841	1845	1849	1854	1858	0	1	1	2	2	3	3	3	4
.27	1862	1866	1871	1875	1879	1884	1888	1892	1897	1901	0	1	1	2	2	3	3	3	4
.28	1905	1910	1914	1919	1923	1928	1932	1936	1941	1945	0	1	1	2	2	3	3	4	4
.29	1950	1954	1959	1963	1968	1972	1977	1982	1986	1991	0	1	1	2	2	3	3	4	4
.30	1995	2000	2004	2009	2014	2018	2023	2028	2032	2037	0	1	1	2	2	3	3	4	4
.31	2042	2046	2051	2056	2061	2065	2070	2075	2080	2084	0	1	1	2	2	3	3	4	4
.32	2089	2094	2099	2104	2109	2113	2118	2123	2128	2133	0	1	1	2	2	3	3	4	4
.33	2138	2143	2148	2153	2158	2163	2168	2173	2178	2183	0	1	1	2	2	3	3	4	4
.34	2188	2193	2198	2203	2208	2213	2218	2223	2228	2234	1	1	2	2	3	3	4	4	5
.35	2239	2244	2249	2254	2259	2265	2270	2275	2280	2286	1	1	2	2	3	3	4	4	5
.36	2291	2296	2301	2307	2312	2317	2323	2328	2333	2339	1	1	2	2	3	3	4	4	5
.37	2344	2350	2355	2360	2366	2371	2377	2382	2388	2393	1	1	2	2	3	3	4	4	5
.38	2399	2404	2410	2415	2421	2427	2432	2438	2443	2449	1	1	2	2	3	3	4	4	5
.39	2455	2460	2466	2472	2477	2483	2489	2495	2500	2506	1	1	2	2	3	3	4	5	5
.40	2512	2518	2523	2529	2535	2541	2547	2553	2559	2564	1	1	2	2	3	4	4	5	5
.41	2570	2576	2582	2588	2594	2600	2606	2612	2618	2624	1	1	2	2	3	4	4	5	5
.42	2630	2636	2642	2649	2655	2661	2667	2673	2679	2685	1	1	2	2	3	4	4	5	6
.43	2692	2698	2704	2710	2716	2723	2729	2735	2742	2748	1	1	2	3	3	4	4	5	6
.44	2754	2761	2767	2773	2780	2786	2793	2799	2805	2812	1	1	2	3	3	4	4	5	6
.45	2818	2825	2831	2838	2844	2851	2858	2864	2871	2877	1	1	2	3	3	4	5	5	6
.46	2884	2891	2897	2904	2911	2917	2924	2931	2938	2944	1	1	2	3	3	4	5	5	6
.47	2951	2958	2965	2972	2979	2985	2992	2999	3006	3013	1	1	2	3	3	4	5	5	6
.48	3020	3027	3034	3041	3048	3055	3062	3069	3076	3083	1	1	2	3	4	4	5	6	6
.49	3090	3097	3105	3112	3119	3126	3133	3141	3148	3155	1	1	2	3	4	4	5	6	6

ANTILOGARITHMS

	0	1	2	3	4	5	6	7	8	9	1	2	3	4	5	6	7	8	9
.50	3162	3170	3177	3184	3192	3199	3206	3214	3221	3228	1	1	2	3	4	4	5	6	7
.51	3236	3243	3251	3258	3266	3273	3281	3289	3296	3304	1	2	2	3	4	5	5	6	7
.52	3311	3319	3327	3334	3342	3350	3357	3365	3373	3381	1	2	2	3	4	5	5	6	7
.53	3388	3396	3404	3412	3420	3428	3436	3443	3451	3459	1	2	2	3	4	5	6	6	7
.54	3467	3475	3483	3491	3499	3508	3516	3524	3532	3540	1	2	2	3	4	5	6	6	7
.55	3548	3556	3565	3573	3581	3589	3597	3606	3614	3622	1	2	2	3	4	5	6	7	7
.56	3631	3639	3648	3656	3664	3673	3681	3690	3698	3707	1	2	3	3	4	5	6	7	8
.57	3715	3724	3733	3741	3750	3758	3767	3776	3784	3793	1	2	3	3	4	5	6	7	8
.58	3802	3811	3819	3828	3837	3846	3855	3864	3873	3882	1	2	3	4	4	5	6	7	8
.59	3890	3899	3908	3917	3926	3936	3945	3954	3963	3972	1	2	3	4	5	5	6	7	8
.60	3981	3990	3999	4009	4018	4027	4036	4046	4055	4064	1	2	3	4	5	6	6	7	8
.61	4074	4083	4093	4102	4111	4121	4130	4140	4150	4159	1	2	3	4	5	6	7	8	9
.62	4169	4178	4188	4198	4207	4217	4227	4236	4246	4256	1	2	3	4	5	6	7	8	9
.63	4266	4276	4285	4295	4305	4315	4325	4335	4345	4355	1	2	3	4	5	6	7	8	9
.64	4365	4375	4385	4395	4406	4416	4426	4436	4446	4457	1	2	3	4	5	6	7	8	9
.65	4467	4477	4487	4498	4508	4519	4529	4539	4550	4560	1	2	3	4	5	6	7	8	9
.66	4571	4581	4592	4603	4613	4624	4634	4645	4656	4667	1	2	3	4	5	6	7	9	10
.67	4677	4688	4699	4710	4721	4732	4742	4753	4764	4775	1	2	3	4	5	7	8	9	10
.68	4786	4797	4808	4819	4831	4842	4853	4864	4875	4887	1	2	3	4	6	7	8	9	10
.69	4898	4909	4920	4932	4943	4955	4966	4977	4989	5000	1	2	3	5	6	7	8	9	10
.70	5012	5023	5035	5047	5058	5070	5082	5093	5105	5117	1	2	4	5	6	7	8	9	11
.71	5129	5140	5152	5164	5176	5188	5200	5212	5224	5236	1	2	4	5	6	7	8	10	11
.72	5248	5260	5272	5284	5297	5309	5321	5333	5346	5358	1	2	4	5	6	7	9	10	11
.73	5370	5383	5395	5408	5420	5433	5445	5458	5470	5483	1	3	4	5	6	8	9	10	11
.74	5495	5508	5521	5534	5546	5559	5572	5585	5598	5610	1	3	4	5	6	8	9	10	12
.75	5623	5636	5649	5662	5675	5689	5702	5715	5728	5741	1	3	4	5	7	8	9	10	12
.76	5754	5768	5781	5794	5808	5821	5834	5848	5861	5875	1	3	4	5	7	8	9	11	12
.77	5888	5902	5916	5929	5943	5957	5970	5984	5998	6012	1	3	4	5	7	8	10	11	12
.78	6026	6039	6053	6067	6081	6095	6109	6124	6138	6152	1	3	4	6	7	8	10	11	13
.79	6166	6180	6194	6209	6223	6237	6252	6266	6281	6295	1	3	4	6	7	9	10	11	13
.80	6310	6324	6339	6353	6368	6383	6397	6412	6427	6442	1	3	4	6	7	9	10	12	13
.81	6457	6471	6486	6501	6516	6531	6546	6561	6577	6592	2	3	5	6	8	9	11	12	14
.82	6607	6622	6637	6653	6668	6683	6699	6714	6730	6745	2	3	5	6	8	9	11	12	14
.83	6761	6776	6792	6808	6823	6839	6855	6871	6887	6902	2	3	5	6	8	9	11	13	14
.84	6918	6934	6950	6966	6982	6998	7015	7031	7047	7063	2	3	5	6	8	10	11	13	15
.85	7079	7096	7112	7129	7145	7161	7178	7194	7211	7228	2	3	5	7	8	10	12	13	15
.86	7244	7261	7278	7295	7311	7328	7345	7362	7379	7396	2	3	5	7	8	10	12	13	15
.87	7413	7430	7447	7464	7482	7499	7516	7534	7551	7568	2	3	5	7	9	10	12	14	16
.88	7586	7603	7621	7638	7656	7674	7691	7709	7727	7745	2	4	5	7	9	11	12	14	16
.89	7762	7780	7798	7816	7834	7852	7870	7889	7907	7925	2	4	5	7	9	11	13	14	16
.90	7943	7962	7980	7998	8017	8035	8054	8072	8091	8110	2	4	6	7	9	11	13	15	17
.91	8128	8147	8166	8185	8204	8222	8241	8260	8279	8299	2	4	6	8	9	11	13	15	17
.92	8318	8337	8356	8375	8395	8414	8433	8453	8472	8492	2	4	6	8	10	12	14	15	17
.93	8511	8531	8551	8570	8590	8610	8630	8650	8670	8690	2	4	6	8	10	12	14	16	18
.94	8710	8730	8750	8770	8790	8810	8831	8851	8872	8892	2	4	6	8	10	12	14	16	18
.95	8913	8933	8954	8974	8995	9016	9036	9057	9078	9099	2	4	6	8	10	12	15	17	19
.96	9120	9141	9162	9183	9204	9226	9247	9268	9290	9311	2	4	6	8	11	13	15	17	19
.97	9333	9354	9376	9397	9419	9441	9462	9484	9506	9528	2	4	7	9	11	13	15	17	20
.98	9550	9572	9594	9616	9638	9661	9683	9705	9727	9750	2	4	7	9	11	13	16	18	20
.99	9772	9795	9817	9840	9863	9886	9908	9931	9954	9977	2	5	7	9	11	14	16	18	20

NATURAL SINES

Degrees	0′ 0°.0	6′ 0°.1	12′ 0°.2	18′ 0°.3	24′ 0°.4	30′ 0°.5	36′ 0°.6	42′ 0°.7	48′ 0°.8	54′ 0°.9	Mean Differences 1	2	3	4	5
0	.0000	0017	0035	0052	0070	0087	0105	0122	0140	0157	3	6	9	12	15
1	.0175	0192	0209	0227	0244	0262	0279	0297	0314	0332	3	6	9	12	15
2	.0349	0366	0384	0401	0419	0436	0454	0471	0488	0506	3	6	9	12	15
3	.0523	0541	0558	0576	0593	0610	0628	0645	0663	0680	3	6	9	12	15
4	.0698	0715	0732	0750	0767	0785	0802	0819	0837	0854	3	6	9	12	15
5	.0872	0889	0906	0924	0941	0958	0976	0993	1011	1028	3	6	9	12	14
6	.1045	1063	1080	1097	1115	1132	1149	1167	1184	1201	3	6	9	12	14
7	.1219	1236	1253	1271	1288	1305	1323	1340	1357	1374	3	6	9	12	14
8	.1392	1409	1426	1444	1461	1478	1495	1513	1530	1547	3	6	9	12	14
9	.1564	1582	1599	1616	1633	1650	1668	1685	1702	1719	3	6	9	12	14
10	.1736	1754	1771	1788	1805	1822	1840	1857	1874	1891	3	6	9	12	14
11	.1908	1925	1942	1959	1977	1994	2011	2028	2045	2062	3	6	9	11	14
12	.2079	2096	2113	2130	2147	2164	2181	2198	2215	2232	3	6	9	11	14
13	.2250	2267	2284	2300	2317	2334	2351	2368	2385	2402	3	6	8	11	14
14	.2419	2436	2453	2470	2487	2504	2521	2538	2554	2571	3	6	8	11	14
15	.2588	2605	2622	2639	2656	2672	2689	2706	2723	2740	3	6	8	11	14
16	.2756	2773	2790	2807	2823	2840	2857	2874	2890	2907	3	6	8	11	14
17	.2924	2940	2957	2974	2990	3007	3024	3040	3057	3074	3	6	8	11	14
18	.3090	3107	3123	3140	3156	3173	3190	3206	3223	3239	3	6	8	11	14
19	.3256	3272	3289	3305	3322	3338	3355	3371	3387	3404	3	5	8	11	14
20	.3420	3437	3453	3469	3486	3502	3518	3535	3551	3567	3	5	8	11	14
21	.3584	3600	3616	3633	3649	3665	3681	3697	3714	3730	3	5	8	11	14
22	.3746	3762	3778	3795	3811	3827	3843	3859	3875	3891	3	5	8	11	14
23	.3907	3923	3939	3955	3971	3987	4003	4019	4035	4051	3	5	8	11	14
24	.4067	4083	4099	4115	4131	4147	4163	4179	4195	4210	3	5	8	11	13
25	.4226	4242	4258	4274	4289	4305	4321	4337	4352	4368	3	5	8	11	13
26	.4384	4399	4415	4431	4446	4462	4478	4493	4509	4524	3	5	8	10	13
27	.4540	4555	4571	4586	4602	4617	4633	4648	4664	4679	3	5	8	10	13
28	.4695	4710	4726	4741	4756	4772	4787	4802	4818	4833	3	5	8	10	13
29	.4848	4863	4879	4894	4909	4924	4939	4955	4970	4985	3	5	8	10	13
30	.5000	5015	5030	5045	5060	5075	5090	5105	5120	5135	3	5	8	10	13
31	.5150	5165	5180	5195	5210	5225	5240	5255	5270	5284	2	5	7	10	12
32	.5299	5314	5329	5344	5358	5373	5388	5402	5417	5432	2	5	7	10	12
33	.5446	5461	5476	5490	5505	5519	5534	5548	5563	5577	2	5	7	10	12
34	.5592	5606	5621	5635	5650	5664	5678	5693	5707	5721	2	5	7	10	12
35	.5736	5750	5764	5779	5793	5807	5821	5835	5850	5864	2	5	7	10	12
36	.5878	5892	5906	5920	5934	5948	5962	5976	5990	6004	2	5	7	9	12
37	.6018	6032	6046	6060	6074	6088	6101	6115	6129	6143	2	5	7	9	12
38	.6157	6170	6184	6198	6211	6225	6239	6252	6266	6280	2	5	7	9	11
39	.6293	6307	6320	6334	6347	6361	6374	6388	6401	6414	2	4	7	9	11
40	.6428	6441	6455	6468	6481	6494	6508	6521	6534	6547	2	4	7	9	11
41	.6561	6574	6587	6600	6613	6626	6639	6652	6665	6678	2	4	7	9	11
42	.6691	6704	6717	6730	6743	6756	6769	6782	6794	6807	2	4	6	9	11
43	.6820	6833	6845	6858	6871	6884	6896	6909	6921	6934	2	4	6	8	11
44	.6947	6959	6972	6984	6997	7009	7022	7034	7046	7059	2	4	6	8	10

NATURAL SINES

Degrees	0' 0°.0	6' 0°.1	12' 0°.2	18' 0°.3	24 0°.4	30' 0°.5	36' 0°.6	42' 0°.7	48' 0°.8	54' 0°.9	Mean Differences 1	2	3	4	5
45	.7071	7083	7096	7108	7120	7133	7145	7157	7169	7181	2	4	6	8	10
46	.7193	7206	7218	7230	7242	7254	7266	7278	7290	7302	2	4	6	8	10
47	.7314	7325	7337	7349	7361	7373	7385	7396	7408	7420	2	4	6	8	10
48	.7431	7443	7455	7466	7478	7490	7501	7513	7524	7536	2	4	6	8	10
49	.7547	7558	7570	7581	7593	7604	7615	7627	7638	7649	2	4	6	8	9
50	.7660	7672	7683	7694	7705	7716	7727	7738	7749	7760	2	4	6	7	9
51	.7771	7782	7793	7804	7815	7826	7837	7848	7859	7869	2	4	5	7	9
52	.7880	7891	7902	7912	7923	7934	7944	7955	7965	7976	2	4	5	7	9
53	.7986	7997	8007	8018	8028	8039	8049	8059	8070	8080	2	3	5	7	9
54	.8090	8100	8111	8121	8131	8141	8151	8161	8171	8181	2	3	5	7	8
55	.8192	8202	8211	8221	8231	8241	8251	8261	8271	8281	2	3	5	7	8
56	.8290	8300	8310	8320	8329	8339	8348	8358	8368	8377	2	3	5	6	8
57	.8387	8396	8406	8415	8425	8434	8443	8453	8462	8471	2	3	5	6	8
58	.8480	8490	8499	8508	8517	8526	8536	8545	8554	8563	2	3	5	6	8
59	.8572	8581	8590	8599	8607	8616	8625	8634	8643	8652	1	3	4	6	7
60	.8660	8669	8678	8686	8695	8704	8712	8721	8729	8738	1	3	4	6	7
61	.8746	8755	8763	8771	8780	8788	8796	8805	8813	8821	1	3	4	6	7
62	.8829	8838	8846	8854	8862	8870	8878	8886	8894	8902	1	3	4	5	7
63	.8910	8918	8926	8934	8942	8949	8957	8965	8973	8980	1	3	4	5	6
64	.8988	8996	9003	9011	9018	9026	9033	9041	9048	9056	1	3	4	5	6
65	.9063	9070	9078	9085	9092	9100	9107	9114	9121	9128	1	2	4	5	6
66	.9135	9143	9150	9157	9164	9171	9178	9184	9191	9198	[illegible]	2	3	5	6
67	.9205	9212	9219	9225	9232	9239	9245	9252	9259	9265	1	2	3	4	6
68	.9272	9278	9285	9291	9298	9304	9311	9317	9323	9330	1	2	3	4	5
69	.9336	9342	9348	9354	9361	9367	9373	9379	9385	9391	1	2	3	4	5
70	.9397	9403	9409	9415	9421	9426	9432	9438	9444	9449	1	2	3	4	5
71	.9455	9461	9466	9472	9478	9483	9489	9494	9500	9505	1	2	3	4	5
72	.9511	9516	9521	9527	9532	9537	9542	9548	9553	9558	1	2	3	3	4
73	.9563	9568	9573	9578	9583	9588	9593	9598	9603	9608	1	2	2	3	4
74	.9613	9617	9622	9627	9632	9636	9641	9646	9650	9655	1	2	2	3	4
75	.9659	9664	9668	9673	9677	9681	9685	9690	9694	9699	1	1	2	3	4
76	.9703	9707	9711	9715	9720	9724	9728	9732	9736	9740	1	1	2	3	3
77	.9744	9748	9751	9755	9759	9763	9767	9770	9774	9778	1	1	2	3	3
78	.9781	9785	9789	9792	9796	9799	9803	9806	9810	9813	1	1	2	2	3
79	.9816	9820	9823	9826	9829	9833	9836	9839	9842	9845	1	1	2	2	3
80	.9848	9851	9854	9857	9860	9863	9866	9869	9871	9874	0	1	1	2	2
81	.9877	9880	9882	9885	9888	9890	9893	9895	9898	9900	0	1	1	2	2
82	.9903	9905	9907	9910	9912	9914	9917	9919	9921	9923	0	1	1	2	2
83	.9925	9928	9930	9932	9934	9936	9938	9940	9942	9943	0	1	1	1	2
84	.9945	9947	9949	9951	9952	9954	9956	9957	9959	9960	0	1	1	1	2
85	.9962	9963	9965	9966	9968	9969	9971	9972	9973	9974	0	0	1	1	1
86	.9976	9977	9978	9979	9980	9981	9982	9983	9984	9985	0	0	1	1	1
87	.9986	9987	9988	9989	9990	9990	9991	9992	9993	9993	0	0	0	1	1
88	.9994	9995	9995	9996	9996	9997	9997	9997	9998	9998	0	0	0	0	0
89	.9998	9999	9999	9999	9999	1.000	1.000	1.000	1.000	1.000	0	0	0	0	0
90	1.000														

NATURAL COSINES

[*Numbers in difference columns to be subtracted, not added*]

Degrees	0′ 0°.0	6′ 0°.1	12′ 0°.2	18′ 0°.3	24′ 0°.4	30′ 0°.5	36′ 0°.6	42′ 0°.7	48′ 0°.8	54′ 0°.9	Mean Differences 1	2	3	4	5
0	1.000	1.000	1.000	1.000	1.000	1.000	.9999	9999	9999	9999	0	0	0	0	0
1	.9998	9998	9998	9997	9997	9997	9996	9996	9995	9995	0	0	0	0	0
2	.9994	9993	9993	9992	9991	9990	9990	9989	9988	9987	0	0	0	1	1
3	.9986	9985	9984	9983	9982	9981	9980	9979	9978	9977	0	0	1	1	1
4	.9976	9974	9973	9972	9971	9969	9968	9966	9965	9963	0	0	1	1	1
5	.9962	9960	9959	9957	9956	9954	9952	9951	9949	9947	0	1	1	1	2
6	.9945	9943	9942	9940	9938	9936	9934	9932	9930	9928	0	1	1	1	2
7	.9925	9923	9921	9919	9917	9914	9912	9910	9907	9905	0	1	1	2	2
8	.9903	9900	9898	9895	9893	9890	9888	9885	9882	9880	0	1	1	2	2
9	.9877	9874	9871	9869	9866	9863	9860	9857	9854	9851	0	1	1	2	2
10	.9848	9845	9842	9839	9836	9833	9829	9826	9823	9820	1	1	2	2	3
11	.9816	9813	9810	9806	9803	9799	9796	9792	9789	9785	1	1	2	2	3
12	.9781	9778	9774	9770	9767	9763	9759	9755	9751	9748	1	1	2	3	3
13	.9744	9740	9736	9732	9728	9724	9720	9715	9711	9707	1	1	2	3	3
14	.9703	9699	9694	9690	9686	9681	9677	9673	9668	9664	1	1	2	3	4
15	.9659	9655	9650	9646	9641	9636	9632	9627	9622	9617	1	2	2	3	4
16	.9613	9608	9603	9598	9593	9588	9583	9578	9573	9568	1	2	2	3	4
17	.9563	9558	9553	9548	9542	9537	9532	9527	9521	9516	1	2	3	3	4
18	.9511	9505	9500	9494	9489	9483	9478	9472	9466	9461	1	2	3	4	5
19	.9455	9449	9444	9438	9432	9426	9421	9415	9409	9403	1	2	3	4	5
20	.9397	9391	9385	9379	9373	9367	9361	9354	9348	9342	1	2	3	4	5
21	.9336	9330	9323	9317	9311	9304	9298	9291	9285	9278	1	2	3	4	5
22	.9272	9265	9259	9252	9245	9239	9232	9225	9219	9212	1	2	3	4	6
23	.9205	9198	9191	9184	9178	9171	9164	9157	9150	9143	1	2	3	5	6
24	.9135	9128	9121	9114	9107	9100	9092	9085	9078	9070	1	2	4	5	6
25	.9063	9056	9048	9041	9033	9026	9018	9011	9003	8996	1	3	4	5	6
26	.8988	8980	8973	8965	8957	8949	8942	8934	8926	8918	1	3	4	5	6
27	.8910	8902	8894	8886	8878	8870	8862	8854	8846	8838	1	3	4	5	7
28	.8829	8821	8813	8805	8796	8788	8780	8771	8763	8755	1	3	4	6	7
29	.8746	8738	8729	8721	8712	8704	8695	8686	8678	8669	1	3	4	6	7
30	.8660	8652	8643	8634	8625	8616	8607	8599	8590	8581	1	3	4	6	7
31	.8572	8563	8554	8545	8536	8526	8517	8508	8499	8490	2	3	5	6	8
32	.8480	8471	8462	8453	8443	8434	8425	8415	8406	8396	2	3	5	6	8
33	.8387	8377	8368	8358	8348	8339	8329	8320	8310	8300	2	3	5	6	8
34	.8290	8281	8271	8261	8251	8241	8231	8221	8211	8202	2	3	5	7	8
35	.8192	8181	8171	8161	8151	8141	8131	8121	8111	8100	2	3	5	7	8
36	.8090	8080	8070	8059	8049	8039	8028	8018	8007	7997	2	3	5	7	9
37	.7986	7976	7965	7955	7944	7934	7923	7912	7902	7891	2	4	5	7	9
38	.7880	7869	7859	7848	7837	7826	7815	7804	7793	7782	2	4	5	7	9
39	.7771	7760	7749	7738	7727	7716	7705	7694	7683	7672	2	4	6	7	9
40	.7660	7649	7638	7627	7615	7604	7593	7581	7570	7559	2	4	6	8	9
41	.7547	7536	7524	7513	7501	7490	7478	7466	7455	7443	2	4	6	8	10
42	.7431	7420	7408	7396	7385	7373	7361	7349	7337	7325	2	4	6	8	10
43	.7314	7302	7290	7278	7266	7254	7242	7230	7218	7206	2	4	6	8	10
44	.7193	7181	7169	7157	7145	7133	7120	7108	7096	7083	2	4	6	8	10

NATURAL COSINES

[Numbers in difference columns to be subtracted, not added]

Degrees	0' 0°.0	6' 0°.1	12' 0°.2	18' 0°.3	24' 0°.4	30' 0°.5	36' 0°.6	42' 0°.7	48' 0°.8	54' 0°.9	Mean Differences 1	2	3	4	5
45	.7071	7059	7046	7034	7022	7009	6997	6984	6972	6959	2	4	6	8	10
46	.6947	6934	6921	6909	6896	6884	6871	6858	6845	6833	2	4	6	8	11
47	.6820	6807	6794	6782	6769	6756	6743	6730	6717	6704	2	4	6	9	11
48	.6691	6678	6665	6652	6639	6626	6613	6600	6587	6574	2	4	7	9	11
49	.6561	6547	6534	6521	6508	6494	6481	6468	6455	6441	2	4	7	9	11
50	.6428	6414	6401	6388	6374	6361	6347	6334	6320	6307	2	4	7	9	11
51	.6293	6280	6266	6252	6239	6225	6211	6198	6184	6170	2	5	7	9	11
52	.6157	6143	6129	6115	6101	6088	6074	6060	6046	6032	2	5	7	9	12
53	.6018	6004	5990	5976	5962	5948	5934	5920	5906	5892	2	5	7	9	12
54	.5878	5864	5850	5835	5821	5807	5793	5779	5764	5750	2	5	7	9	12
55	.5736	5721	5707	5693	5678	5664	5650	5635	5621	5606	2	5	7	10	12
56	.5592	5577	5563	5548	5534	5519	5505	5490	5476	5461	2	5	7	10	12
57	.5446	5432	5417	5402	5388	5373	5358	5344	5329	5314	2	5	7	10	12
58	.5299	5284	5270	5255	5240	5225	5210	5195	5180	5165	2	5	7	10	12
59	.5150	5135	5120	5105	5090	5075	5060	5045	5030	5015	3	5	8	10	13
60	.5000	4985	4970	4955	4939	4924	4909	4894	4879	4863	3	5	8	10	13
61	.4848	4833	4818	4802	4787	4772	4756	4741	4726	4710	3	5	8	10	13
62	.4695	4679	4664	4648	4633	4617	4602	4586	4571	4555	3	5	8	10	13
63	.4540	4524	4509	4493	4478	4462	4446	4431	4415	4399	3	5	8	10	13
64	.4384	4368	4352	4337	4321	4305	4289	4274	4258	4242	3	5	8	11	13
65	.4226	4210	4195	4179	4163	4147	4131	4115	4099	4083	3	5	8	11	13
66	.4067	4051	4035	4019	4003	3987	3971	3955	3939	3923	3	5	8	11	14
67	.3907	3891	3875	3859	3843	3827	3811	3795	3778	3762	3	5	8	11	14
68	.3746	3730	3714	3697	3681	3665	3649	3633	3616	3600	3	5	8	11	14
69	.3584	3567	3551	3535	3518	3502	3486	3469	3453	3437	3	5	8	11	14
70	.3420	3404	3387	3371	3355	3338	3322	3305	3289	3272	3	5	8	11	14
71	.3256	3239	3223	3206	3190	3173	3156	3140	3123	3107	3	6	8	11	14
72	.3090	3074	3057	3040	3024	3007	2990	2974	2957	2940	3	6	8	11	14
73	.2924	2907	2890	2874	2857	2840	2823	2807	2790	2773	3	6	8	11	14
74	.2756	2740	2723	2706	2689	2672	2656	2639	2622	2605	3	6	8	11	14
75	.2588	2571	2554	2538	2521	2504	2487	2470	2453	2436	3	6	8	11	14
76	.2419	2402	2385	2368	2351	2334	2317	2300	2284	2267	3	6	8	11	14
77	.2250	2233	2215	2198	2181	2164	2147	2130	2113	2096	3	6	9	11	14
78	.2079	2062	2045	2028	2011	1994	1977	1959	1942	1925	3	6	9	11	14
79	.1908	1891	1874	1857	1840	1822	1805	1788	1771	1754	3	6	9	11	14
80	.1736	1719	1702	1685	1668	1650	1633	1616	1599	1582	3	6	9	12	14
81	.1564	1547	1530	1513	1495	1478	1461	1444	1426	1409	3	6	9	12	14
82	.1392	1374	1357	1340	1323	1305	1288	1271	1253	1236	3	6	9	12	14
83	.1219	1201	1184	1167	1149	1132	1115	1097	1080	1063	3	6	9	12	14
84	.1045	1028	1011	0993	0976	0958	0941	0924	0906	0889	3	6	9	12	14
85	.0872	0854	0837	0819	0802	0785	0767	0750	0732	0715	3	6	9	12	15
86	.0698	0680	0663	0645	0628	0610	0593	0576	0558	0541	3	6	9	12	15
87	.0523	0506	0488	0471	0454	0436	0419	0401	0384	0366	3	6	9	12	15
88	.0349	0332	0314	0297	0279	0262	0244	0227	0209	0192	3	6	9	12	15
89	.0175	0157	0140	0122	0105	0087	0070	0052	0035	0017	3	6	9	12	15
90	.0000														

NATURAL TANGENTS

Degrees	0' 0°.0	6' 0°.1	12' 0°.2	18' 0°.3	24' 0°.4	30' 0°.5	36' 0°.6	42' 0°.7	48' 0°.8	54' 0°.9	Mean Differences 1	2	3	4	5
0	.0000	0017	0035	0052	0070	0087	0105	0122	0140	0157	3	6	9	12	15
1	.0175	0192	0209	0227	0244	0262	0279	0297	0314	0332	3	6	9	12	15
2	.0349	0367	0384	0402	0419	0437	0454	0472	0489	0507	3	6	9	12	15
3	.0524	0542	0559	0577	0594	0612	0629	0647	0664	0682	3	6	9	12	15
4	.0699	0717	0734	0752	0769	0787	0805	0822	0840	0857	3	6	9	12	15
5	.0875	0892	0910	0928	0945	0963	0981	0998	1016	1033	3	6	9	12	15
6	.1051	1069	1086	1104	1122	1139	1157	1175	1192	1210	3	6	9	12	15
7	.1228	1246	1263	1281	1299	1317	1334	1352	1370	1388	3	6	9	12	15
8	.1405	1423	1441	1459	1477	1495	1512	1530	1548	1566	3	6	9	12	15
9	.1584	1602	1620	1638	1655	1673	1691	1709	1727	1745	3	6	9	12	15
10	.1763	1781	1799	1817	1835	1853	1871	1890	1908	1926	3	6	9	12	15
11	.1944	1962	1980	1998	2016	2035	2053	2071	2089	2107	3	6	9	12	15
12	.2126	2144	2162	2180	2199	2217	2235	2254	2272	2290	3	6	9	12	15
13	.2309	2327	2345	2364	2382	2401	2419	2438	2456	2475	3	6	9	12	15
14	.2493	2512	2530	2549	2568	2586	2605	2623	2642	2661	3	6	9	12	16
15	.2679	2698	2717	2736	2754	2773	2792	2811	2830	2849	3	6	9	13	16
16	.2867	2886	2905	2924	2943	2962	2981	3000	3019	3038	3	6	9	13	16
17	.3057	3076	3096	3115	3134	3153	3172	3191	3211	3230	3	6	10	13	16
18	.3249	3269	3288	3307	3327	3346	3365	3385	3404	3424	3	6	10	13	16
19	.3443	3463	3482	3502	3522	3541	3561	3581	3600	3620	3	7	10	13	16
20	.3640	3659	3679	3699	3719	3739	3759	3779	3799	3819	3	7	10	13	17
21	.3839	3859	3879	3899	3919	3939	3959	3979	4000	4020	3	7	10	13	17
22	.4040	4061	4081	4101	4122	4142	4163	4183	4204	4224	3	7	10	14	17
23	.4245	4265	4286	4307	4327	4348	4369	4390	4411	4431	3	7	10	14	17
24	.4452	4473	4494	4515	4536	4557	4578	4599	4621	4642	4	7	11	14	18
25	.4663	4684	4706	4727	4748	4770	4791	4813	4834	4856	4	7	11	14	18
26	.4877	4899	4921	4942	4964	4986	5008	5029	5051	5073	4	7	11	15	18
27	.5095	5117	5139	5161	5184	5206	5228	5250	5272	5295	4	7	11	15	18
28	.5317	5340	5362	5384	5407	5430	5452	5475	5498	5520	4	8	11	15	19
29	.5543	5566	5589	5612	5635	5658	5681	5704	5727	5750	4	8	12	15	19
30	.5774	5797	5820	5844	5867	5890	5914	5938	5961	5985	4	8	12	16	20
31	.6009	6032	6056	6080	6104	6128	6152	6176	6200	6224	4	8	12	16	20
32	.6249	6273	6297	6322	6346	6371	6395	6420	6445	6469	4	8	12	16	20
33	.6494	6519	6544	6569	6594	6619	6644	6669	6694	6720	4	8	13	17	21
34	.6745	6771	6796	6822	6847	6873	6899	6924	6950	6976	4	9	13	17	21
35	.7002	7028	7054	7080	7107	7133	7159	7186	7212	7239	4	9	13	18	22
36	.7265	7292	7319	7346	7373	7400	7427	7454	7481	7508	5	9	14	18	23
37	.7536	7563	7590	7618	7646	7673	7701	7729	7757	7785	5	9	14	18	23
38	.7813	7841	7869	7898	7926	7954	7983	8012	8040	8069	5	9	14	19	24
39	.8098	8127	8156	8185	8214	8243	8273	8302	8332	8361	5	10	15	20	24
40	.8391	8421	8451	8481	8511	8541	8571	8601	8632	8662	5	10	15	20	25
41	.8693	8724	8754	8785	8816	8847	8878	8910	8941	8972	5	10	16	21	26
42	.9004	9036	9067	9099	9131	9163	9195	9228	9260	9293	5	11	16	21	27
43	.9325	9358	9391	9424	9457	9490	9523	9556	9590	9623	6	11	17	22	28
44	.9657	9691	9725	9759	9793	9827	9861	9896	9930	9965	6	11	17	23	29

NATURAL TANGENTS

Degrees	0' 0°.0	6' 0°.1	12' 0°.2	18' 0°.3	24' 0°.4	30' 0°.5	36' 0°.6	42' 0°.7	48' 0°.8	54' 0°.9	Mean Differences 1	2	3	4	5
45	1.0000	0035	0070	0105	0141	0176	0212	0247	0283	0319	6	12	18	24	30
46	1.0355	0392	0428	0464	0501	0538	0575	0612	0649	0686	6	12	18	25	31
47	1.0724	0761	0799	0837	0875	0913	0951	0990	1028	1067	6	13	19	25	32
48	1.1106	1145	1184	1224	1263	1303	1343	1383	1423	1463	7	13	20	27	33
49	1.1504	1544	1585	1626	1667	1708	1750	1792	1833	1875	7	14	21	28	34
50	1.1918	1960	2002	2045	2088	2131	2174	2218	2261	2305	7	14	22	29	36
51	1.2349	2393	2437	2482	2527	2572	2617	2662	2708	2753	8	15	23	30	38
52	1.2799	2846	2892	2938	2985	3032	3079	3127	3175	3222	8	16	24	31	39
53	1.3270	3319	3367	3416	3465	3514	3564	3613	3663	3713	8	16	25	33	41
54	1.3764	3814	3865	3916	3968	4019	4071	4124	4176	4229	9	17	26	34	43
55	1.4281	4335	4388	4442	4496	4550	4605	4659	4715	4770	9	18	27	36	45
56	1.4826	4882	4938	4994	5051	5108	5166	5224	5282	5340	10	19	29	38	48
57	1.5399	5458	5517	5577	5637	5697	5757	5818	5880	5941	10	20	30	40	50
58	1.6003	6066	6128	6191	6255	6319	6383	6447	6512	6577	11	21	32	43	53
59	1.6643	6709	6775	6842	6909	6977	7045	7113	7182	7251	11	23	34	45	56
60	1.7321	7391	7461	7532	7603	7675	7747	7820	7893	7966	12	24	36	48	60
61	1.8040	8115	8190	8265	8341	8418	8495	8572	8650	8728	13	26	38	51	64
62	1.8807	8887	8967	9047	9128	9210	9292	9375	9458	9542	14	27	41	55	68
63	1.9626	9711	9797	9883	9970	2.0057	2.0145	2.0233	2.0323	2.0413	15	29	44	58	73
64	2.0503	0594	0686	0778	0872	0965	1060	1155	1251	1348	16	31	47	63	78
65	2.1445	1543	1642	1742	1842	1943	2045	2148	2251	2355	17	34	51	68	85
66	2.2460	2566	2673	2781	2889	2998	3109	3220	3332	3445	18	37	55	73	92
67	2.3559	3673	3789	3906	4023	4142	4262	4383	4504	4627	20	40	60	79	99
68	2.4751	4876	5002	5129	5257	5386	5517	5649	5782	5916	22	43	65	87	108
69	2.6051	6187	6325	6464	6605	6746	6889	7034	7179	7326	24	47	71	95	119
70	2.7475	7625	7776	7929	8083	8239	8397	8556	8716	8878	26	52	78	104	131
71	2.9042	9208	9375	9544	9714	9887	3.0061	3.0237	3.0415	3.0595	29	58	87	116	145
72	3.0777	0961	1146	1334	1524	1716	1910	2106	2305	2506	32	64	96	129	161
73	3.2709	2914	3122	3332	3544	3759	3977	4197	4420	4646	36	72	108	144	180
74	3.4874	5105	5339	5576	5816	6059	6305	6554	6806	7062	41	81	122	163	204
75	3.7321	7583	7848	8118	8391	8667	8947	9232	9520	9812	46	93	139	186	232
76	4.0108	0408	0713	1022	1335	1653	1976	2303	2635	2972	53	107	160	213	267
77	4.3315	3662	4015	4374	4737	5107	5483	5864	6252	6646					
78	4.7046	7453	7867	8288	8716	9152	9594	5.0045	5.0504	5.0970	Mean differences cease to be sufficiently accurate.				
79	5.1446	1929	2422	2924	3435	3955	4486	5026	5578	6140					
80	5.6713	7297	7894	8502	9124	9758	6.0405	6.1066	6.1742	6.2432					
81	6.3138	3859	4596	5350	6122	6912	7720	8548	9395	7.0264					
82	7.1154	2066	3002	3962	4947	5958	6996	8062	9158	8.0285					
83	8.1443	2636	3863	5126	6427	7769	9152	9.0579	9.2052	9.3572					
84	9.5144	9.677	9.845	10.02	10.20	10.39	10.58	10.78	10.99	11.20					
85	11.43	11.66	11.91	12.16	12.43	12.71	13.00	13.30	13.62	13.95					
86	14.30	14.67	15.06	15.46	15.89	16.35	16.83	17.34	17.89	18.46					
87	19.08	19.74	20.45	21.20	22.02	22.90	23.86	24.90	26.03	27.27					
88	28.64	30.14	31.82	33.69	35.80	38.19	40.92	44.07	47.74	52.08					
89	57.29	63.66	71.62	81.85	95.49	114.6	143.2	191.0	286.5	573.0					
90	∞														

LOGARITHMS OF SINES

Degrees	0' 0°.0	6' 0°.1	12' 0°.2	18' 0°.3	24' 0°.4	30' 0°.5	36' 0°.6	42' 0°.7	48' 0°.8	54' 0°.9	Mean Differences 1	2	3	4	5
0	$-\infty$	$\bar{3}$.2419	$\bar{3}$.5429	7190	8439	9408	$\bar{2}$.0200	$\bar{2}$.0870	$\bar{2}$.1450	$\bar{2}$.1961					
1	$\bar{2}$.2419	2832	3210	3558	3880	4179	4459	4723	4971	5206					
2	$\bar{2}$.5428	5640	5842	6035	6220	6397	6567	6731	6889	7041					
3	$\bar{2}$.7188	7330	7468	7602	7731	7857	7979	8098	8213	8326					
4	$\bar{2}$.8436	8543	8647	8749	8849	8946	9042	9135	9226	9315	16	32	48	64	80
5	$\bar{2}$.9403	9489	9573	9655	9736	9816	9894	9970	$\bar{1}$.0046	$\bar{1}$.0120	13	26	39	52	65
6	$\bar{1}$.0192	0264	0334	0403	0472	0539	0605	0670	0734	0797	11	22	33	44	55
7	$\bar{1}$.0859	0920	0981	1040	1099	1157	1214	1271	1326	1381	10	19	29	38	48
8	$\bar{1}$.1436	1489	1542	1594	1646	1697	1747	1797	1847	1895	8	17	25	34	42
9	$\bar{1}$.1943	1991	2038	2085	2131	2176	2221	2266	2310	2353	8	15	23	30	38
10	$\bar{1}$.2397	2439	2482	2524	2565	2606	2647	2687	2727	2767	7	14	20	27	34
11	$\bar{1}$.2806	2845	2883	2921	2959	2997	3034	3070	3107	3143	6	12	19	25	31
12	$\bar{1}$.3179	3214	3250	3284	3319	3353	3387	3421	3455	3488	6	11	17	23	28
13	$\bar{1}$.3521	3554	3586	3618	3650	3682	3713	3745	3775	3806	5	11	16	21	26
14	$\bar{1}$.3837	3867	3897	3927	3957	3986	4015	4044	4073	4102	5	10	15	20	2[illegible]
15	$\bar{1}$.4130	4158	4186	4214	4242	4269	4296	4323	4350	4377	5	9	14	18	23
16	$\bar{1}$.4403	4430	4456	4482	4508	4533	4559	4584	4609	4634	4	9	13	17	21
17	$\bar{1}$.4659	4684	4709	4733	4757	4781	4805	4829	4853	4876	4	8	12	16	20
18	$\bar{1}$.4900	4923	4946	4969	4992	5015	5037	5060	5082	5104	4	8	11	15	19
19	$\bar{1}$.5126	5148	5170	5192	5213	5235	5256	5278	5299	5320	4	7	11	14	18
20	$\bar{1}$.5341	5361	5382	5402	5423	5443	5463	5484	5504	5523	3	7	10	14	17
21	$\bar{1}$.5543	5563	5583	5602	5621	5641	5660	5679	5698	5717	3	6	10	13	16
22	$\bar{1}$.5736	5754	5773	5792	5810	5828	5847	5865	5883	5901	3	6	9	12	15
23	$\bar{1}$.5919	5937	5954	5972	5990	6007	6024	6042	6059	6076	3	6	9	12	15
24	$\bar{1}$.6093	6110	6127	6144	6161	6177	6194	6210	6227	6243	3	6	8	11	14
25	$\bar{1}$.6259	6276	6292	6308	6324	6340	6356	6371	6387	6403	3	5	8	11	13
26	$\bar{1}$.6418	6434	6449	6465	6480	6495	6510	6526	6541	6556	3	5	8	10	13
27	$\bar{1}$.6570	6585	6600	6615	6629	6644	6659	6673	6687	6702	2	5	7	10	12
28	$\bar{1}$.6716	6730	6744	6759	6773	6787	6801	6814	6828	6842	2	5	7	9	12
29	$\bar{1}$.6856	6869	6883	6896	6910	6923	6937	6950	6963	6977	2	4	7	9	11
30	$\bar{1}$.6990	7003	7016	7029	7042	7055	7068	7080	7093	7106	2	4	6	9	11
31	$\bar{1}$.7118	7131	7144	7156	7168	7181	7193	7205	7218	7230	2	4	6	8	10
32	$\bar{1}$.7242	7254	7266	7278	7290	7302	7314	7326	7338	7349	2	4	6	8	10
33	$\bar{1}$.7361	7373	7384	7396	7407	7419	7430	7442	7453	7464	2	4	6	8	10
34	$\bar{1}$.7476	7487	7498	7509	7520	7531	7542	7553	7564	7575	2	4	6	7	9
35	$\bar{1}$.7586	7597	7607	7618	7629	7640	7650	7661	7671	7682	2	4	5	7	9
36	$\bar{1}$.7692	7703	7713	7723	7734	7744	7754	7764	7774	7785	2	3	5	7	9
37	$\bar{1}$.7795	7805	7815	7825	[illegible]35	7844	7854	7864	7874	7884	2	3	5	7	8
38	$\bar{1}$.7893	7903	7913	7922	7932	7941	7951	7960	7970	7979	2	3	5	6	8
39	$\bar{1}$.7989	7998	8007	8017	8026	8035	8044	8053	8063	8072	2	3	5	6	8
40	$\bar{1}$.8081	8090	8099	8108	8117	8125	8134	8143	8152	8161	1	3	4	6	7
41	$\bar{1}$.8169	8178	8187	8195	8204	8213	8221	8230	8238	8247	1	3	4	6	7
42	$\bar{1}$.8255	8264	8272	8280	8289	8297	8305	8313	8322	8330	1	3	4	6	7
43	$\bar{1}$.8338	8346	8354	8362	8370	8378	8396	8394	8402	8410	1	3	4	5	7
44	$\bar{1}$.8418	8426	8433	8441	8449	8457	8464	8472	8480	8487	1	3	4	5	6

LOGARITHMS OF SINES

Degrees	0′ 0°.0	6′ 0°.1	12′ 0°.2	18′ 0°.3	24′ 0°.4	30′ 0°.5	36′ 0°.6	42′ 0°.7	48′ 0°.8	54′ 0°.9	Mean Differences 1	2	3	4	5
45	$\bar{1}$.8495	8502	8510	8517	8525	8532	8540	8547	8555	8562	1	2	4	5	6
46	$\bar{1}$.8569	8577	8584	8591	8598	8606	8613	8620	8627	8634	1	2	4	5	6
47	$\bar{1}$.8641	8648	8655	8662	8669	8676	8683	8690	8697	8704	1	2	3	5	6
48	$\bar{1}$.8711	8718	8724	8731	8738	8745	8751	8758	8765	8771	1	2	3	4	6
49	$\bar{1}$.8778	8784	8791	8797	8804	8810	8817	8823	8830	8836	1	2	3	4	5
50	$\bar{1}$.8843	8849	8855	8862	8868	8874	8880	8887	8893	8899	1	2	3	4	5
51	$\bar{1}$.8905	8911	8917	8923	8929	8935	8941	8947	8953	8959	1	2	3	4	5
52	$\bar{1}$.8965	8971	8977	8983	8989	8995	9000	9006	9012	9018	1	2	3	4	5
53	$\bar{1}$.9023	9029	9035	9041	9046	9052	9057	9063	9069	9074	1	2	3	4	5
54	$\bar{1}$.9080	9085	9091	9096	9101	9107	9112	9118	9123	9128	1	2	3	4	5
55	$\bar{1}$.9134	9139	9144	9149	9155	9160	9165	9170	9175	9181	1	2	3	3	4
56	$\bar{1}$.9186	9191	9196	9201	9206	9211	9216	9221	9226	9231	1	2	3	3	4
57	$\bar{1}$.9236	9241	9246	9251	9255	9260	9265	9270	9275	9279	1	2	2	3	4
58	$\bar{1}$.9284	9289	9294	9298	9303	9308	9312	9317	9322	9326	1	2	2	3	4
59	$\bar{1}$.9331	9335	9340	9344	9349	9353	9358	9362	9367	9371	1	1	2	3	4
60	$\bar{1}$.9375	9380	9384	9388	9393	9397	9401	9406	9410	9414	1	1	2	3	4
61	$\bar{1}$.9418	9422	9427	9431	9435	9439	9443	9447	9451	9455	1	1	2	3	3
62	$\bar{1}$.9459	9463	9467	9471	9475	9479	9483	9487	9491	9495	1	1	2	3	3
63	$\bar{1}$.9499	9503	9506	9510	9514	9518	9522	9525	9529	9533	1	1	2	3	3
64	$\bar{1}$.9537	9540	9544	9548	9551	9555	9558	9562	9566	9569	1	1	2	2	3
65	$\bar{1}$.9573	9576	9580	9583	9587	9590	9594	9597	9601	9604	1	1	2	2	3
66	$\bar{1}$.9607	9611	9614	9617	9621	9624	9627	9631	9634	9637	1	1	2	2	3
67	$\bar{1}$.9640	9643	9647	9650	9653	9656	9659	9662	9666	9669	1	1	2	2	3
68	$\bar{1}$.9672	9675	9678	9681	9684	9687	9690	9693	9696	9699	0	1	1	2	2
69	$\bar{1}$.9702	9704	9707	9710	9713	9716	9719	9722	9724	9727	0	1	1	2	2
70	$\bar{1}$.9730	9733	9735	9738	9741	9743	9746	9749	9751	9754	0	1	1	2	2
71	$\bar{1}$.9757	9759	9762	9764	9767	9770	9772	9775	9777	9780	0	1	1	2	2
72	$\bar{1}$.9782	9785	9787	9789	9792	9794	9797	9799	9801	9804	0	1	1	2	2
73	$\bar{1}$.9806	9808	9811	9813	9815	9817	9820	9822	9824	9826	0	1	1	2	2
74	$\bar{1}$.9828	9831	9833	9835	9837	9839	9841	9843	9845	9847	0	1	1	1	2
75	$\bar{1}$.9849	9851	9853	9855	9857	9859	9861	9863	9865	9867	0	1	1	1	2
76	$\bar{1}$.9869	9871	9873	9875	9876	9878	9880	9882	9884	9885	0	1	1	1	2
77	$\bar{1}$.9887	9889	9891	9892	9894	9896	9897	9899	9901	9902	0	1	1	1	1
78	$\bar{1}$.9904	9906	9907	9909	9910	9912	9913	9915	9916	9918	0	1	1	1	1
79	$\bar{1}$.9919	9921	9922	9924	9925	9927	9928	9929	9931	9932	0	0	1	1	1
80	$\bar{1}$.9934	9935	9936	9937	9939	9940	9941	9943	9944	9945	0	0	1	1	1
81	$\bar{1}$.9946	9947	9949	9950	9951	9952	9953	9954	9955	9956	0	0	1	1	1
82	$\bar{1}$.9958	9959	9960	9961	9962	9963	9964	9965	9966	9967	0	0	1	1	1
83	$\bar{1}$.9968	9968	9969	9970	9971	9972	9973	9974	9975	9975	0	0	0	1	1
84	$\bar{1}$.9976	9977	9978	9978	9979	9980	9981	9981	9982	9983	0	0	0	0	1
85	$\bar{1}$.9983	9984	9985	9985	9986	9987	9987	9988	9988	9989	0	0	0	0	0
86	$\bar{1}$.9989	9990	9990	9991	9991	9992	9992	9993	9993	9994	0	0	0	0	0
87	$\bar{1}$.9994	9994	9995	9995	9996	9996	9996	9997	9997	9997	0	0	0	0	0
88	$\bar{1}$.9997	9998	9998	9998	9998	9999	9999	9999	9999	9999	0	0	0	0	0
89	$\bar{1}$.9999	9999	0.0000	0000	0000	0000	0000	0000	0000	0000					
90	0.0000														

LOGARITHMS OF COSINES

[*Numbers in difference columns to be subtracted, not added*]

Degrees	0′ 0°.0	6′ 0°.1	12′ 0°.2	18′ 0°.3	24′ 0°.4	30′ 0°.5	36′ 0°.6	42′ 0°.7	48′ 0°.8	54′ 0°.9	Mean Differences 1	2	3	4	5
0	0.0000	0000	0000	0000	0000	0000	0000	0000	0000	$\bar{1}$.9999	0	0	0	0	0
1	$\bar{1}$.9999	9999	9999	9999	9999	9999	9998	9998	9998	9998	0	0	0	0	0
2	$\bar{1}$.9997	9997	9997	9997	9996	9996	9996	9995	9995	9994	0	0	0	0	0
3	$\bar{1}$.9994	9994	9993	9993	9992	9992	9991	9991	9990	9990	0	0	0	0	0
4	$\bar{1}$.9989	9989	9988	9988	9987	9987	9986	9985	9985	9984	0	0	0	0	0
5	$\bar{1}$.9983	9983	9982	9981	9981	9980	9979	9978	9978	9977	0	0	0	0	1
6	$\bar{1}$.9976	9975	9975	9974	9973	9972	9971	9970	9969	9968	0	0	0	1	1
7	$\bar{1}$.9968	9967	9966	9965	9964	9963	9962	9961	9960	9959	0	0	1	1	1
8	$\bar{1}$.9958	9956	9955	9954	9953	9952	9951	9950	9949	9947	0	0	1	1	1
9	$\bar{1}$.9946	9945	9944	9943	9941	9940	9939	9937	9936	9935	0	0	1	1	1
10	$\bar{1}$.9934	9932	9931	9929	9928	9927	9925	9924	9922	9921	0	0	1	1	1
11	$\bar{1}$.9919	9918	9916	9915	9913	9912	9910	9909	9907	9906	0	1	1	1	1
12	$\bar{1}$.9904	9902	9901	9899	9897	9896	9894	9892	9891	9889	0	[illegible]	1	1	1
13	$\bar{1}$.9887	9885	9884	9882	9880	9878	9876	9875	9873	9871	0	1	1	1	2
14	$\bar{1}$.9869	9867	9865	9863	9861	9859	9857	9855	9853	9851	0	1	1	1	2
15	$\bar{1}$.9849	9847	9845	9843	9841	9839	9837	9835	9833	9831	0	1	1	1	2
16	$\bar{1}$.9828	9826	9824	9822	9820	9817	9815	9813	9811	9808	0	1	1	2	2
17	$\bar{1}$.9806	9804	9801	9799	9797	9794	9792	9789	9787	9785	0	1	1	2	2
18	$\bar{1}$.9782	9780	9777	9775	9772	9770	9767	9764	9762	9759	0	1	1	2	2
19	$\bar{1}$.9757	9754	9751	9749	9746	9743	9741	9738	9735	9733	0	1	1	2	2
20	$\bar{1}$.9730	9727	9724	9722	9719	9716	9713	9710	9707	9704	0	1	1	2	2
21	$\bar{1}$.9702	9699	9696	9693	9690	9687	9684	9681	9678	9675	0	1	1	2	2
22	$\bar{1}$.9672	9669	9666	9662	9659	9656	9653	9650	9647	9643	1	1	2	2	3
23	$\bar{1}$.9640	9637	9634	9631	9627	9624	9621	9617	9614	9611	1	1	2	2	3
24	$\bar{1}$.9607	9604	9601	9597	9594	9590	9587	9583	9580	9576	1	1	2	2	3
25	$\bar{1}$.9573	9569	9566	9562	9558	9555	9551	9548	9544	9540	1	1	2	2	3
26	$\bar{1}$.9537	9533	9529	9525	9522	9518	9514	9510	9507	9503	1	1	2	3	3
27	$\bar{1}$.9499	9495	9491	9487	9483	9479	9475	9471	9467	9463	1	1	2	3	3
28	$\bar{1}$.9459	9455	9451	9447	9443	9439	9435	9431	9427	9422	1	1	2	3	3
29	$\bar{1}$.9418	9414	9410	9406	9401	9397	9393	9388	9384	9380	1	1	2	3	4
30	$\bar{1}$.9375	9371	9367	9362	9358	9353	9349	9344	9340	9335	1	1	2	3	4
31	$\bar{1}$.9331	9326	9322	9317	9312	9308	9303	9298	9294	9289	1	2	2	3	4
32	$\bar{1}$.9284	9279	9275	9270	9265	9260	9255	9251	9246	9241	1	2	2	3	4
33	$\bar{1}$.9236	9231	9226	9221	9216	9211	9206	9201	9196	9191	1	2	3	3	4
34	$\bar{1}$.9186	9181	9175	9170	9165	9160	9155	9149	9144	9139	1	2	3	3	4
35	$\bar{1}$.9134	9128	9123	9118	9112	9107	9101	9096	9091	9085	1	2	3	4	5
36	$\bar{1}$.9080	9074	9069	9063	9057	9052	9046	9041	9035	9029	1	2	3	4	5
37	$\bar{1}$.9023	9018	9012	9006	9000	8995	8989	8983	8977	8971	1	2	3	4	5
38	$\bar{1}$.8965	8959	8953	8947	8941	8935	8929	8923	8917	8911	1	2	3	4	5
39	$\bar{1}$.8905	8899	8893	8887	8880	8874	8868	8862	8855	8849	1	2	3	4	5
40	$\bar{1}$.8843	8836	8830	8823	8817	8810	8804	8797	8791	8784	1	2	3	4	5
41	$\bar{1}$.8778	8771	8765	8758	8751	8745	8738	8731	8724	8718	1	2	3	5	6
42	$\bar{1}$.8711	8704	8697	8690	8683	8676	8669	8662	8655	8648	1	2	3	5	6
43	$\bar{1}$.8641	8634	8627	8620	8613	8606	8598	8591	8584	8577	1	2	4	5	6
44	$\bar{1}$.8569	8562	[illegible]	8547	8540	8532	8525	8517	8510	8502	1	2	4	5	6

LOGARITHMS OF COSINES

[Numbers in difference columns to be subtracted, not added]

Degrees	0' 0°.0	6' 0°.1	12' 0°.2	18' 0°.3	24' 0°.4	30' 0°.5	36' 0°.6	42' 0°.7	48' 0°.8	54' 0°.9	Mean Differences 1	2	3	4	5
45	$\bar{1}$.8495	8487	8480	8472	8464	8457	8449	8441	8433	8426	1	3	4	5	6
46	$\bar{1}$.8418	8410	8402	8394	8386	8378	8370	8362	8354	8346	1	3	4	5	7
47	$\bar{1}$.8338	8330	8322	8313	8305	8297	8289	8280	8272	8264	1	3	4	6	7
48	$\bar{1}$.8255	8247	8238	8230	8221	8213	8204	8195	8187	8178	1	3	4	6	7
49	$\bar{1}$.8169	8161	8152	8143	8134	8125	8117	8108	8099	8090	1	3	4	6	7
50	$\bar{1}$.8081	8072	8063	8053	8044	8035	8026	8017	8007	7998	2	3	5	6	8
51	$\bar{1}$.7989	7979	7970	7960	7951	7941	7932	7922	7913	7903	2	3	5	6	8
52	$\bar{1}$.7893	7884	7874	7864	7854	7844	7835	7825	7815	7805	2	3	5	7	8
53	$\bar{1}$.7795	7785	7774	7764	7754	7744	7734	7723	7713	7703	2	3	5	7	9
54	$\bar{1}$.7692	7682	7671	7661	7650	7640	7629	7618	7607	7597	2	4	5	7	9
55	$\bar{1}$.7586	7575	7564	7553	7542	7531	7520	7509	7498	7487	2	4	6	7	9
56	$\bar{1}$.7476	7464	7453	7442	7430	7419	7407	7396	7384	7373	2	4	6	8	10
57	$\bar{1}$.7361	7349	7338	7326	7314	7302	7290	7278	7266	7254	2	4	6	8	10
58	$\bar{1}$.7242	7230	7218	7205	7193	7181	7168	7156	7144	7131	2	4	6	8	10
59	$\bar{1}$.7118	7106	7093	7080	7068	7055	7042	7029	7016	7003	2	4	6	9	11
60	$\bar{1}$.6990	6977	6963	6950	6937	6923	6910	6896	6883	6869	2	4	7	9	11
61	$\bar{1}$.6856	6842	6828	6814	6801	6787	6773	6759	6744	6730	2	5	7	9	12
62	$\bar{1}$.6716	6702	6687	6673	6659	6644	6629	6615	6600	6585	2	5	7	10	12
63	$\bar{1}$.6570	6556	6541	6526	6510	6495	6480	6465	6449	6434	3	5	8	10	13
64	$\bar{1}$.6418	6403	6387	6371	6356	6340	6324	6308	6292	6276	3	5	8	11	13
65	$\bar{1}$.6259	6243	6227	6210	6194	6177	6161	6144	6127	6110	3	6	8	11	14
66	$\bar{1}$.6093	6076	6059	6042	6024	6007	5990	5972	5954	5937	3	6	9	12	15
67	$\bar{1}$.5919	5901	5883	5865	5847	5828	5810	5792	5773	5754	3	6	9	12	15
68	$\bar{1}$.5736	5717	5698	5679	5660	5641	5621	5602	5583	5563	3	6	10	13	16
69	$\bar{1}$.5543	5523	5504	5484	5463	5443	5423	5402	5382	5361	3	7	10	14	17
70	$\bar{1}$.5341	5320	5299	5278	5256	5235	5213	5192	5170	5148	4	7	11	14	18
71	$\bar{1}$.5126	5104	5082	5060	5037	5015	4992	4969	4946	4923	4	8	11	15	19
72	$\bar{1}$.4900	4876	4853	4829	4805	4781	4757	4733	4709	4684	4	8	12	16	20
73	$\bar{1}$.4659	4634	4609	4584	4559	4533	4508	4482	4456	4430	4	9	13	17	21
74	$\bar{1}$.4403	4377	4350	4323	4296	4269	4242	4214	4186	4158	5	9	14	18	23
75	$\bar{1}$.4130	4102	4073	4044	4015	3986	3957	3927	3897	3867	5	10	15	20	24
76	$\bar{1}$.3837	3806	3775	3745	3713	3682	3650	3618	3586	3554	5	11	16	21	26
77	$\bar{1}$.3521	3488	3455	3421	3387	3353	3319	3284	3250	3214	6	11	17	23	28
78	$\bar{1}$.3179	3143	3107	3070	3034	2997	2959	2921	2883	2845	6	12	19	25	31
79	$\bar{1}$.2806	2767	2727	2687	2647	2606	2565	2524	2482	2439	7	14	20	27	34
80	$\bar{1}$.2397	2353	2310	2266	2221	2176	2131	2085	2038	1991	8	15	23	30	38
81	$\bar{1}$.1943	1895	1847	1797	1747	1697	1646	1594	1542	1489	8	17	25	34	42
82	$\bar{1}$.1436	1381	1326	1271	1214	1157	1099	1040	0981	0920	10	19	29	38	48
83	$\bar{1}$.0859	0797	0734	0670	0605	0539	0472	0403	0334	0264	11	22	33	44	55
84	$\bar{1}$.0192	0120	0046	$\bar{2}$.9970	$\bar{2}$.9894	$\bar{2}$.9816	$\bar{2}$.9736	$\bar{2}$.9655	$\bar{2}$.9573	$\bar{2}$.9489	13	26	39	52	65
85	$\bar{2}$.9403	9315	9226	9135	9042	8946	8849	8749	8647	8543	16	32	48	64	80
86	$\bar{2}$.8436	8326	8213	8098	7979	7857	7731	7602	7468	7330					
87	$\bar{2}$.7188	7041	6889	6731	6567	6397	6220	6035	5842	5640					
88	$\bar{2}$.5428	5206	4971	4723	4459	4179	3880	3558	3210	2832					
89	$\bar{2}$.2419	1961	1450	0870	0200	$\bar{3}$.9408	$\bar{3}$.8439	$\bar{3}$.7190	$\bar{3}$.5429	$\bar{3}$.2419					
90	$-\infty$														